Energy, Environment, and Sustainability

AIMS AND SCOPE

This books series publishes cutting edge monographs and professional books focused on all aspects of energy and environmental sustainability, especially as it relates to energy concerns. The Series is published in partnership with the International Society for Energy, Environment, and Sustainability. The books in these series are edited or authored by top researchers and professional across the globe. The series aims at publishing state-of-the-art research and development in areas including, but not limited to:

- Renewable Energy
- Alternative Fuels
- Engines and Locomotives
- Combustion and Propulsion
- Fossil Fuels
- Carbon Capture
- Control and Automation for Energy
- Environmental Pollution
- Waste Management
- Transportation Sustainability

Review Process

The proposal for each volume is reviewed by the main editor and/or the advisory board. The chapters in each volume are individually reviewed single blind by expert reviewers (at least four reviews per chapter) and the main editor.

Ethics Statement for this series can be found in the Springer standard guidelines here https://www.springer.com/us/authors-editors/journal-author/journal-author-helpdesk/before-you-start/before-you-start/1330#c14214

Burak Zincir · Pravesh Chandra Shukla ·
Avinash Kumar Agarwal
Editors

Decarbonization of Maritime Transport

Springer

Editors
Burak Zincir
Department of Marine Engineering
Istanbul Technical University
Istanbul, Türkiye

Pravesh Chandra Shukla
Department of Mechanical Engineering
Indian Institute of Technology Bhilai
Raipur, India

Avinash Kumar Agarwal
Department of Mechanical Engineering
Indian Institute of Technology Kanpur
Kanpur, India

ISSN 2522-8366 ISSN 2522-8374 (electronic)
Energy, Environment, and Sustainability
ISBN 978-981-99-1676-4 ISBN 978-981-99-1677-1 (eBook)
https://doi.org/10.1007/978-981-99-1677-1

This Springer imprint is published by the registered company Springer Nature Singapore Pte Ltd.
The registered company address is: 152 Beach Road, #21-01/04 Gateway East, Singapore 189721, Singapore

Preface

The Secretary-General of the International Maritime Organization (IMO), an agency of the United Nations for regulating international shipping, said, "Decarbonizing international shipping is a priority issue for IMO, and we are all committed to acting together in achieving the highest possible ambition" at the Second IMO Symposium on Low- and Zero-carbon Fuels for Shipping on October 21, 2022. With this spirit, this book consists of various decarbonization methods for maritime transport from different perspectives of the contributing authors to act together.

The International Society for Energy, Environment and Sustainability (ISEES) was founded at the Indian Institute of Technology Kanpur (IIT Kanpur), India, in January 2014, to spread knowledge/awareness and catalyze research activities in the fields of energy, environment, sustainability, and combustion. The society's goal is to contribute to the development of clean, affordable, and secure energy resources and a sustainable environment for society and spread knowledge in the areas mentioned above, and create awareness about the environmental challenges the world is facing today. The unique way adopted by ISEES was to break the conventional silos of specializations (engineering, science, environment, agriculture, biotechnology, materials, fuels, etc.) to tackle the problems related to energy, environment, and sustainability in a holistic manner. This is quite evident in the participation of experts from all fields to resolve these issues. The ISEES is involved in various activities, such as conducting workshops, seminars, and conferences, in the domains of its interests. The society also recognizes the outstanding works of young scientists, professionals, and engineers for their contributions in these fields by conferring them awards under various categories.

Sixth International Conference on "Sustainable Energy and Environmental Challenges" (VI-SEEC) was organized under the auspices of ISEES from December 27 to 29, 2021, in hybrid mode due to restrictions on travel because of the ongoing COVID-19 pandemic situation. This conference provided a platform for discussions between eminent scientists and engineers from various countries, including India, Spain, Austria, Australia, South Korea, Brazil, Mexico, USA, Malaysia, Japan, Hong

Kong, China, the UK, Netherlands, Poland, Finland, Italy, Israel, Kenya, Türkiye, and Saudi Arabia. At this conference, eminent international speakers presented their views on energy, combustion, emissions, and alternative energy resources for sustainable development and a cleaner environment. The conference presented two high-voltage plenary talks by Prof. Ashutosh Sharma, Secretary, DST and Dr. V. K. Saraswat, Honorable Member, NITI Ayog.

The conference included 12 technical panel discussions on energy and environmental sustainability topics. Each session had 6–7 eminent scientists who shared their opinion and discussed the trends for the future. The technical sessions at the conference included Fuels for Sustainable Transport, Challenges for Desalination and Wastewater Treatment and Possible Solutions, Engine Combustion Modelling, Simulation and Sprays, Bioenergy/biofuels, Coal Biomass Combustion for Power Generation, Microbial Processes and Products, Future of IC Engine Technology and Roadmap, Air Pollution and Climate Change: Sustainable Approaches, Sustainable Energy from Carbon Neutral Sources, Biological Waste Treatment, Combustion: Emerging Paradigm, and Thermochemical Processes for Biomass. A total of 500+ participants and speakers from around the world attended this three-day conference.

This conference laid out the roadmap for technology development, opportunities, and challenges in energy, environment, and sustainability domains. All these topics are very relevant for the country and the world in the present context. We acknowledge the support from various agencies and organizations for conducting the Sixth ISEES conference (VI-SEEC), where these books germinated. We want to acknowledge our publishing partner Springer (special thanks to Ms. Swati Mehershi).

The editors would like to express their sincere gratitude to many authors worldwide for submitting their high-quality work on time and revising it appropriately at short notice. We want to express our special gratitude to our prolific set of reviewers, Prof. Elen Twrdy, Prof. Cengiz Deniz, Dr. Giacomo Belgiorno, Dr. Gabriele Di Blasio, Dr. Caglar Dere, Dr. Levent Bilgili, Dr. Maja Stojaković, Mr. Bugra Arda Zincir, and Dr. Omer Berkehan Inal, who reviewed various chapters of this monograph and provided their valuable suggestions to improve the manuscripts.

This book provides an overview of the methods for decarbonization of maritime transport. The book includes a life cycle analysis of alternative marine fuels, hydrogen-fueled marine engines, investigation of alternative marine fuels for ships, electrification and hybridization of ferries, carbon capture systems for ships, green port concept studies, energy efficiency and management implementation, and evaluation of the market-based measures for maritime transport to achieve decarbonization. Chapters include recent results and focus on current trends in the maritime transport sector. In this book, readers will understand various decarbonization methods in different areas, such as shipboard applications, ports, and the maritime transport market. Some of the studies are the review of the state of the art, and some are analyses that show the possible ways to decarbonize maritime transport effectively. We hope the book will greatly interest the professionals and post-graduate students involved in fuels, internal combustion engines, the maritime industry, and environmental research.

Istanbul, Türkiye — Burak Zincir
Bhilai, India — Pravesh Chandra Shukla
Kanpur, India — Avinash Kumar Agarwal

Contents

About the Editors

Dr. Burak Zincir has been working at the Marine Engineering Department of Maritime Faculty, Istanbul Technical University, since 2013, and he has been Assistant Professor since 2020 at the same department. His research areas are alternative marine fuels, performance and emissions of marine diesel engines, decarbonization methods and technologies, and energy efficiency on ships. His M.Sc. thesis was on applying hydrogen as marine diesel engine fuel. His Ph.D. thesis focused on an alternative fuel assessment model for ships and experiments on the effect of methanol on diesel engines. He did the experimental part of his Ph.D. study at Lund University, Sweden, in 2018. Istanbul Technical University awarded him "The Most Successful Dissertation Thesis Award" in 2019. He works on alternative marine fuels such as ammonia, biofuels, and electrofuels.

Dr. Pravesh Chandra Shukla is Assistant Professor in the Department of Mechanical Engineering at Indian Institute of Technology (IIT) Bhilai, India. Dr. Shukla received his Ph.D. from IIT Kanpur and has also worked as Senior Research Associate with the institute. He was Postdoctoral Researcher in the Division of Combustion Engines, Department of Energy Sciences, Lund University, Sweden. He briefly worked in Ecole Centrale de Nantes, France, in the field of dual fuel combustion. He is Recipient of Young Scientist Award from the International Society for Energy, Environment and Sustainability. Dr. Shukla mainly works in the field of internal combustion engines and alternative fuels for transportation. He worked on the development of additives for high compression ratio heavy-duty engines fueled with alcohol. He is involved in investigating the emission characteristics for alternative fuels like biodiesel, HVO, and alcohols for conventional and advanced heavy-duty compression ignition engines. During his doctoral, he was mainly involved in physicochemical characterization of diesel engine exhaust using non-noble metal-based mixed oxides diesel oxidation catalysts. Till now, he has published more than 30 technical articles in international journals and conference proceedings.

Prof. Avinash Kumar Agarwal joined Indian Institute of Technology (IIT) Kanpur in 2001. He worked at the Engine Research Center, UW@Madison, USA, as Post-doctoral Fellow (1999–2001). His interests are IC engines, combustion, alternate and conventional fuels, lubricating oil tribology, optical diagnostics, laser ignition, HCCI, emissions, and particulate control, 1D and 3D simulations of engine processes, and large-bore engines. Professor Agarwal has published 495+ peer-reviewed international journal and conference papers, 70 edited books, 129 books chapters, and 14460+ Scopus and 22300+ Google Scholar citations. Professor Agarwal is Fellow of SAE (2012), Fellow of ASME (2013), Fellow of ISEES (2015), Fellow of INAE (2015), Fellow of NASI (2018), Fellow of Royal Society of Chemistry (2018), and Fellow of American Association of Advancement in Science (2020). He is the recipient of several prestigious awards such as Clarivate Analytics India Citation Award-2017 in Engineering and Technology, NASI-Reliance Industries Platinum Jubilee Award-2012; INAE Silver Jubilee Young Engineer Award-2012; Dr. C. V. Raman Young Teachers Award: 2011; SAE Ralph R. Teetor Educational Award-2008; INSA Young Scientist Award-2007; UICT Young Scientist Award-2007; and INAE Young Engineer Award-2005. Professor Agarwal received Prestigious CSIR Shanti Swarup Bhatnagar Award-2016 in Engineering Sciences. Professor Agarwal is conferred upon Sir J. C. Bose National Fellowship (2019) by SERB for his outstanding contributions. Professor Agarwal was a highly cited researcher (2018) and was in the top ten HCR from India among 4000 HCR researchers globally in 22 fields of inquiry. Also, he has been nominated for Member of the Council of Indian Institute of Technology by the President of India for three years (2022–2025).

Part I
General

Chapter 1
Introduction to Decarbonization of Maritime Transport

Burak Zincir, Pravesh Chandra Shukla, and Avinash Kumar Agarwal

Abstract Maritime transport is the most important mode of transport. Ninety per cent of world trade is carried out by sea transportation. Although maritime transport is the cleanest type of transport considering the amount of carbon dioxide released per tonne transported, some measures must be taken to comply with the Paris Agreement on Climate Change. The International Maritime Organization, which sets the rules for international maritime transport, has conducted many studies to control and reduce emissions from ships. Studies on CO_2 have recently been accelerated, and the IMO Initial Greenhouse Gas (GHG) Strategy was announced in 2018. One of the objectives of this strategy is to reduce CO_2 emissions by 40% in 2030 and 70% in 2050 compared to 2008. The other aim is to reduce greenhouse gas emissions by 50% in 2050 compared to 2008. This strategy, announced by IMO, was the first study in which the maritime sector complied with the Paris Agreement on Climate Change. To achieve these goals, IMO has identified short-term, mid-term, and long-term candidate measures in this strategy and left it to the maritime transport stakeholders to use one or more of them on their ships. This book in the series is called "Decarbonization of Maritime Transport" and includes studies on decarbonization in shipping. The book consists of 10 sections apart from this section which is the introduction section. The book chapters are selected from studies on candidate measures announced in the IMO Initial GHG Strategy. The book includes studies on alternative fuels, carbon capture technology, green port studies, energy efficiency applications on ships, and market-based measures.

Keywords Decarbonization · Alternative fuels · Life cycle analysis · Energy management · Maritime transport

B. Zincir (✉)
Department of Marine Engineering, Istanbul Technical University, Istanbul, Türkiye
e-mail: bzincir@itu.edu.tr

P. C. Shukla
Department of Mechanical Engineering, Indian Institute of Technology Bhilai, Raipur, India

A. K. Agarwal
Department of Mechanical Engineering, Indian Institute of Technology Kanpur, Kanpur, India

B. Zincir et al. (eds.), *Decarbonization of Maritime Transport*, Energy, Environment, and Sustainability, https://doi.org/10.1007/978-981-99-1677-1_1

1.1 Introduction

Maritime transport is the most important mode of transport. Ninety percent of world trade is carried out by sea transportation (Deniz and Zincir 2016). According to the United Nations Conference on Trade and Development (UNCTAD) data, in 2021, there were 99,800 ships of 100 gross tonnes and above in international maritime transport (UNCTAD 2021). The vast majority of these ships use poor-quality fuel, HFO. Although maritime transport is the cleanest type of transport considering the amount of carbon dioxide released per tonne transported, some measures must be taken to comply with the Paris Agreement on Climate Change. The International Maritime Organization, which sets the rules for international maritime transport, has conducted many studies to control and reduce emissions from ships. Limits have been set for nitrogen oxides (NO_x), sulfur oxides (SO_x), and particulate matter (PM) released by ships. Some studies have been carried out to reduce carbon dioxide (CO_2) emissions, and measures have been put forward for both newly built and existing ships since 2011. Studies on CO_2 have recently been accelerated, and the IMO Initial Greenhouse Gas (GHG) Strategy was announced in 2018. One of the objectives of this strategy is to reduce CO_2 emissions by 40% in 2030 and 70% in 2050 compared to 2008. The other aim is to reduce greenhouse gas emissions by 50% in 2050 compared to 2008 (IMO 2023). This strategy, announced by IMO, was the first study in which the maritime sector complied with the Paris Agreement on Climate Change (ABS 2020). To achieve these goals, IMO has identified short-term, mid-term, and long-term candidate measures in this strategy and left it to the maritime transport stakeholders to use one or more of them on their ships.

Along with the IMO Initial GHG Strategy, the term decarbonization in maritime transport has increased its popularity and has been one of the focal points of the studies. This book in the series is called "Decarbonization of Maritime Transport" and includes studies on decarbonization in shipping. The book consists of 10 sections apart from this section which is the introduction section. The book chapters are selected from studies on candidate measures announced in the IMO Initial GHG Strategy. The book includes studies on alternative fuels, carbon capture technology, green port studies, energy efficiency applications on ships, and market-based measures.

Chapter 2, "LCA Studies on Marine Alternative Fuels", includes the life cycle analysis of alternative fuels (ammonia, hydrogen, liquefied natural gas (LNG), methanol, etc.) that can be used in maritime transport. In the chapter, the studies that IMO has done on CO_2 emissions so far are discussed, and the IMO Initial GHG Strategy is mentioned. Life cycle studies of alternative marine fuels in the literature have been examined. As a result of the study, it has been stated that there has been an increase in studies related to alternative fuels since 2000. Although using alternative fuels is a good way to reach the IMO's 2030 and 2050 targets, the current lack of technological infrastructure seems to be an obstacle. Increasing alternative fuel production capacity and supply points is important in replacing traditional fuels with alternative fuels. In addition, it has been stated that focusing on the life cycle studies of alternative fuels will more realistically highlight the real environmental contributions of these fuels.

Chapter 3, "Lifecycle Emissions of Fossil Fuels and Biofuels for Maritime Transportation: A Requirement Analysis", includes the comparison of biofuels with conventional marine fuels in terms of the life cycle, which can be used as a transition fuel in the period of a complete transition to alternative fuels in marine transportation. In this section, firstly, the IMO Initial GHG Strategy is mentioned. In the next step of the study, the role of biofuels in maritime transport is revealed based on the current age distribution of the world fleet. It has been stated that the production methods of biofuels are classified under three generations, and different production methods have been put forward. What life cycle means, the life cycle in fossil fuels and the life cycle in biofuels are explained. The last step of the study compared the life cycle approaches of traditional fossil fuels and biofuels. It has been stated that biofuels produced differently have different life cycle evaluations. It has been explained that biofuels will contribute to maritime transport compared to traditional fossil fuels if fuels are assessed according to their life cycle impact.

Chapter 4, "Hydrogen Fuelled Engine Technology, Adaptation, and Application for Marine Engines", focuses on using carbon-free hydrogen fuel in marine diesel engines. After explaining the physicochemical properties of hydrogen and comparing it with other fuels, the combustion methods of hydrogen in internal combustion engines are explained. The advantages and disadvantages of hydrogen in internal combustion engines are stated and discussed in its application to marine engines. Although hydrogen has an advantage in decarbonization, it is not expected to use neat hydrogen in marine diesel engines in the near future.

Chapter 5, "Investigation and Examination of LNG, Methanol, and Ammonia Usage on Marine Vessels", includes determining the strengths and weaknesses of alternative fuels with SWOT analysis and then comparing these fuels with the TOPSIS method using various criteria. The study focuses on LNG, methanol, and ammonia as alternative marine fuels. While applying the TOPSIS method, expert opinions were used. In the last step of the study, sensitivity analysis was performed, and the effect of changes in criteria scoring on the results was examined. As a result, ammonia, LNG, and methanol ranking were obtained.

Chapter 6, "Electrification and Hybridization of Ferries: State of the Art and Case study", covers the electrification and hybridization of coastal ferries and passenger ships. The first part of the chapter explains fully electric and hybrid propulsion topologies and power management strategies. In the second part of the chapter, a ship sailing in the Bosporus Strait is considered a case ship. Mathematical models with electric and hybrid propulsion systems are explained according to the voyage profile.

Chapter 7, "SWOT Analysis of Carbon Capture, Storage and Transportation for Maritime Industry", reviews carbon capture technologies and their application to ships. In the study, the strengths and weaknesses, opportunities and threats of carbon capture technologies to be used on ships were examined using SWOT analysis. In the study, firstly, the IMO Initial GHG Strategy is mentioned, and why CCS is a system that can be used for maritime transport is explained. Carbon capture technologies and methods are explained in detail, and then the use of CCS on ships is analysed by SWOT analysis. The weak sides of the system are that the system to be installed on the ship is not mature, has a complex structure, will create an extra load for the ship's crew, and the possibility of reducing the cargo carrying capacity. As a result of the study, it was stated that although CCS has some weaknesses, it can be one of the important methods to achieve the IMO's 2030 and 2050 targets.

Chapter 8, "Green Concepts of Ports and Transition Model", focuses on ports to decarbonize maritime transport. In this context, a new approach called the green concept framework was created, and the ports used in the study were evaluated with this approach with 13 criteria. In addition, the deficiencies in the ports were determined according to ISO 50001 principles. At the end of the study, recommendations were made regarding what to do to ensure energy and environmental sustainability.

Chapter 9, "Evaluation of the Green Port Concept for Decarbonized Maritime Industry", focuses on the green port concept as in the previous chapter. The content of the study includes the availability of new green technologies, the availability of low-carbon or zero-carbon terminal equipment, and the development of other energy-consuming elements to achieve the green port concept. In the study, the priorities of the EU ports for the last five years are discussed, and the importance of the green port concept is mentioned depending on these priorities. ISO standards related to the green port concept and the studies done by IMO are explained. In the later part of the chapter, it has been examined what improvements can be made regarding both port operations and ships arriving at the port in terms of green port applications.

Chapter 10, "Energy Efficiency and Management Onboard Ships", contains the energy management framework to increase the operational efficiency of a tanker selected as the case ship for this study. In the chapter, the ISO 50001 energy management standard is explained, and its connection with the applications in the maritime sector is explained. Then, the energy management framework created in the study based on ISO 50001 is introduced, and its application on the selected case ship is explained. As a result of the study, it is stated that ISO 50001-based energy management systems can be adapted to maritime transport.

Chapter 11, "Evaluation of the Market-based Measures by the IMO Criteria: Effects of Current Dynamics", begins with the history of market-based measures (MBMs) and which measures are at the forefront. In the study's next step, MBMs in the literature were compared. While making the comparison, nine criteria were used in creating MBMs by adapting them to today's conditions. As a result of the study, it was determined that the bunker levy met seven criteria, while the other important MBM, ETS, met only three criteria.

The chapters in the book show that there is no single way to decarbonize maritime transport. One or more measures should be used to achieve the 2030 and 2050 targets the IMO set. Each measure has its strengths and weaknesses, according to what is stated in the chapters. Maritime industry stakeholders should try to achieve IMO's goals by choosing the ones suitable for their fleet or infrastructure from these measures. Achieving IMO's goals will also mean reaching the Paris Agreement on Climate Change goals.

Specific chapters covered in the book include:

- Introduction to Decarbonization of Maritime Transport
- LCA Studies on Marine Alternative Fuels
- Lifecycle Emissions of Fossil Fuels and Biofuels for Maritime Transportation: A Requirement Analysis
- Hydrogen Fuelled Engine Technology, Adaptation, and Application for Marine Engines
- Investigation and Examination of LNG, Methanol, and Ammonia Usage on Marine Vessels
- Electrification and Hybridization of Ferries: State of the Art and Case study
- SWOT Analysis of Carbon Capture, Storage, and Transportation for the Maritime Industry
- Green Concept of Ports and Transition Model
- Evaluation of the Green Port Concept for Decarbonized Maritime Industry
- Energy Efficiency and Management Onboard Ships
- Evaluation of the Market-based Measures by the IMO Criteria: Effects of Current Dynamics.

References

American Bureau of Shipping (ABS) (2020) Setting the course to low carbon shipping – pathways to sustainable shipping.

Deniz C, Zincir B (2016) Environmental and economical assessment of alternative marine fuels. J Clean Prod 296:438–449. https://doi.org/10.1016/j.clepro.2015.11.089

International Maritime Organization (IMO) (2023) Initial IMO GHG Strategy. https://www.imo.org/en/MediaCentre/HotTopics/Pages/Reducing-greenhouse-gas-emissions-from-ships.aspx

United Nations Conference on Trade and Development (UNCTAD) (2021) Review of Maritime Transport 2021

Part II
Alternative Marine Fuels

Chapter 2
LCA Studies on Marine Alternative Fuels

Levent Bilgili

Abstract Maritime decarbonization is considered one of the most current and important issues in the maritime industry. The studies on the subject constitute a strong infrastructure for the rules and regulations that the International Maritime Organization should put forward as a top authority. Among these studies, papers and reports examining the subject from a life cycle perspective are of particular importance. Examining the environmental impacts of alternative fuels used on ships with a holistic approach from the cradle-to-grave perspective is very important to determine which method is "really" environmentally friendly. Life cycle assessment not only allows methods to be compared but also explains which process needs to be environmentally corrected among all the processes of the method. Within the scope of this study, first, basic information on the impacts of shipping to the climate change and life cycle assessment was presented, and then, alternative fuels (e.g., ammonia, hydrogen, liquefied natural gas, methanol, etc.) are examined from a life cycle perspective. Thus, it is aimed to evaluate the applications of life cycle assessment in the maritime sector holistically.

Keywords Life cycle assessment · Greenhouse gas · Alternative fuels · Maritime energy

Abbreviations

CH_4	Methane
CII	Carbon Intensity Indicator
CO_2	Carbon Dioxide
CO_2eq	Carbon Dioxide Equivalent
CO	Carbon Monoxide
DME	Dimethyl Ether

L. Bilgili (✉)
Department of Naval Architecture, Maritime Faculty, Bandirma Onyedi Eylul University, Bandirma, Balikesir 10200, Türkiye
e-mail: lbilgili@bandirma.edu.tr

B. Zincir et al. (eds.), *Decarbonization of Maritime Transport*, Energy, Environment, and Sustainability, https://doi.org/10.1007/978-981-99-1677-1_2

ECA	Emission Control Area
EEDI	Energy Efficiency Design Index
EEOI	Energy Efficiency Operation Indicator
EEXI	Energy Efficiency Existing Ship Index
F-T	Fischer–Tropsch
GHGs	Greenhouse gases
GT	Gross Tonnage
GTL	Gas-to-Liquid
GWP	Global Warming Potential
H_2	Hydrogen
HFO	Heavy Fuel Oil
ICE	Internal Combustion Engine
IMO	International Maritime Organization
IPCC	Intergovernmental Panel on Climate Change
LBG	Liquefied Biogas
LCA	Life Cycle Assessment
LH_2	Liquefied Hydrogen
LNG	Liquefied Natural Gas
LPG	Liquefied Petroleum Gas
MARPOL	International Convention for the Prevention of Pollution from Ships
MDO	Marine Diesel Oil
MEPC	Marine Environment Protection Committee
MGO	Marine Gas Oil
NO_x	Nitrogen Oxides
PM	Particulate Matter
SCR	Selective Catalytic Reduction
SDGs	Sustainable Development Goals
SEEMP	Ship Energy Efficiency Management Plan
SO_x	Sulphur Oxides
SOFC	Solid Oxide Fuel Cell
TEU	Twenty-feet Equivalent
UN	United Nations
VOCs	Volatile Organic Compounds

2.1 Introduction

The global demand for energy, which is the foundation of human civilization, has been on a steady increase from the past to the present, and as of 2018, the total global supply is at the level of 14,282 Mtoe. Only 18.7% of this amount is obtained from sources other than fossil-based energy sources, and oil is still the largest resource with a share of 31.6%. 6.8% of this amount is consumed by the maritime sector (IEA 2020). Based on this consumption, ship operations are responsible for the greenhouse gas (GHG) production of 1,076 million tons of carbon dioxide equivalent (CO_2eq) in 2018, which corresponds to an increase of 9.3% compared to 2012. Ships are also responsible for 2.89% of the total global production of anthropogenic emissions (Faber et al. 2020). Although the effects of the COVID-19 pandemic have not yet been determined, the amount of cargo transported by sea increased by 425% between 1970 and 2019 (UNCTAD 2020), and despite the 31.4% increase in the world's CO_2 production between 1971 and 2015, the future looks bleak and dark considering the 76.8% increase in the CO_2 production of maritime industry (IEA 2017). According to a study examining different scenarios, it is predicted that even in the most optimistic scenario, ship-related fuel consumption will increase by 43.5% in 2050 (Eyring et al. 2050). Accordingly, ship-related emissions in 2050 are expected to increase by 90–130% compared to 2008 and 50% compared to 2018 (Faber et al. 2020). In a recent study, it was stated that if the ships were a country, they would be in the sixth or seventh place in the ranking of the countries producing the most CO_2 (IRENA 2021). Besides, in the Baltic Sea, shipping activities are estimated to cause €2.9 billion, which consists of the damage values for marine eutrophication, marine ecotoxicity, air quality, and climate change (Ytreberg et al. 2021).

According to the latest Intergovernmental Panel on Climate Change (IPCC) report, global warming emerges as a very urgent problem that requires immediate action (IPCC 2021), and the United Nations (UNs) have declared the period between 2021 and 2030 as the Ocean Decade of Ocean Science for Sustainable Development (Intergovernmental Oceanographic Commission 2021). In addition to these regulations within the scope of the International Convention for the Prevention of Pollution from Ships (MARPOL), International Maritime Organization (IMO) has also conducted extensive studies on ship-related GHGs. GHGs first entered the agenda of IMO in 1997, and necessary studies were initiated with the Resolution 8 text of the Marine Environment Protection Committee (MEPC). Then, in 2011, Energy Efficiency Design Index (EEDI), Energy Efficiency Operational Indicator (EEOI), and Ship Energy Efficiency Management Plan (SEEMP) came into effect. IMO has published four reports on ship-related GHGs in 2000, 2009, 2014, and 2020, each updating the previous one (Bilgili 2021; Serra and Fancello 2020). At the 76th MEPC meeting held in June 2021, Energy Efficiency Existing Ship Index (EEXI) and Carbon Intensity Indicator (CII) were developed in addition to EEDI.

These rules have been developed and updated so that ships can operate more efficiently and can be compared with each other in terms of energy efficiency. EEDI and EEOI are indicators of CO_2 production of ships at the design stage and during a single operation period, respectively. The calculation of these values is carried out through multivariate equations. Thus, a ship has a score while it is being built and different indicator values for each operation (depending on the load it carries). EEXI and CII are the updated and improved versions of EEDI and EEOI values.

The Initial IMO Strategy on Reduction of GHG Emissions from Ships, published in 2018, outlines IMO's short, medium, and long-term goals. These goals have been determined in accordance with the UN's Sustainable Development Goals (SDGs) and the UN's 2030 Agenda for Sustainable Development, as stated in the reference (United Nations 2022), and within the scope of "urgent measures for climate change and its effects" in article 13 of this agenda, it is aimed to reduce ship-related CO_2 emissions by 40% by 2030 and by 70% by 2050 (IMO 2018; IMO 2022). The short (2018–2023), medium (2023–2030), and long-term (after 2030) targets of IMO on energy efficiency and emission reduction are presented in Table 2.1. These studies are in line with the Paris Agreement, which was prepared on December 12, 2015, and sets the framework for the methods adopted by most countries to ensure that the global average temperature does not exceed 1.5 °C (UNFCCC 2015).

Life cycle assessment (LCA) is an innovative approach used in the holistic study of the environmental, economic, and social impacts of a product system. LCA is an effective method used to obtain realistic results in the calculation of cumulative effects as well as in determining which process is more harmful or beneficial in environmental, economic, and social terms by examining all aspects of the product system.

In this study, LCA studies on alternative fuels used on ships were comprehensively evaluated, and future predictions were made. The aim of the study is to make a general compilation and evaluation of LCA studies on alternative marine fuels and to discuss the compatibility of these studies with IMO and UN targets. Thus, it is aimed to make

Table 2.1 IMO initial strategies (adopted from IMO 2018)

2018–2023	– Developing the existing energy efficiency framework with a focus on EEDI and SEEMP,
	– Developing technical and operational energy efficiency measures for new and existing ships,
	– Assessment of methane (CH_4) and volatile organic compounds (VOCs) emissions,
	– Developing national action plans and policies for the reduction of GHGs in accordance with IMO rules,
	– Using shore-side electricity from renewable sources, establishing infrastructure for alternative low and zero-carbon fuels,
	– Establishing research and development units covering ship propulsion, alternative low and zero-carbon fuels and innovative technologies to increase ship energy efficiency,
	– Increasing life cycle GHG studies for alternative low and zero-carbon fuels,
	– Further studies on emission reduction cost and alternative low and zero-carbon fuels
2023–2030	– Developing an application program for the efficient use of alternative low and zero-carbon fuels,
	– Implementing operational energy efficiency measures for new and existing ships,
	– Developing new and innovative emission reduction mechanisms
2030–	– Following the development of zero-carbon fuels to assess decarbonization in the second half of the twenty-first century,
	– Providing incentives for the development of other new and innovative emission reduction mechanisms

a general analysis of the current situation by creating a basis for the studies to be carried out on the subject in the coming years.

2.2 Summary of the LCA Studies on Alternative Marine Fuels

Making fuels globally acceptable is all about considering them from a life cycle perspective, beyond technological, environmental, economic, and social approaches. It is difficult to accept the fuel as an alternative if it causes high emissions in the production and supply process or if large agricultural lands are needed for production (Gilbert et al. 2018). Or, (so-called) carbon-free alternative fuels may only shift the carbon emissions elsewhere in the supply chain, instead of cutting them off, totally (Wang and Wright 2021). Therefore, the life cycle perspective, in which the production, distribution, and utilization processes of products are examined holistically, is a very important issue for the calculation of the cumulative environmental burden of fuels. The studies are put in order chronologically so that the development can be observed.

In a study in which a 20 kW solid oxide fuel cell (SOFC) system running on renewable methanol was compared with natural gas/biogas and hydrogen-powered batteries from the life cycle perspective, it was concluded that the most gas was formed as a result of the use of biogas, and the use of methanol/biomethanol caused the greatest environmental damage, especially in terms of global warming (Strazza et al. 2010).

In a study comparing the environmental effects of the use of liquefied natural gas (LNG), scrubber, and selective catalytic reduction (SCR) from a life cycle perspective, 1 tkm transportation of a Ro-Ro ship was determined as the functional unit. Although the use of LNG provides a significant reduction in nitrogen oxide (NO_x) and sulfur oxide (SO_x) emissions, a slight reduction in global warming potential (GWP) due to methane leakage was observed (Bengtsson et al. 2011).

In a comprehensive thesis conducted in 2013, the environmental performances of heavy fuel oil (HFO), marine diesel oil/marine gas oil (MDO/MGO), LNG, methanol, dimethyl ether (DME), and Fischer–Tropsch (F-T) diesel fuels were compared in the life cycle perspective. In the study, mainly agricultural land use, GWP, and formation of particulate matter (PM) were examined. One year of operation was used as the functional unit. According to the results, while the use of LNG significantly reduces PM formation, it does not offer a change in terms of GWP. The use of methanol, DME, and F-T diesel reduces the GWP by 56%, 80%, and 78%, respectively. Agricultural land use is increasing in fuels of organic origin. Final results show that F-T diesel and DME are the most environmentally friendly fuels (Øberg 2013).

The aim of another thesis conducted in 2014 is to compare the environmental performances of different alternative fuels with the LCA method. In the study, HFO, MGO, gas-to-liquid (GTL), biodiesels obtained from rapeseed oil and willow, LNG,

LBGs obtained from natural gas, agricultural wastes, and forest wastes, methanol obtained from natural gas and biomass were compared. According to the results, the least energy consumption was realized by LNG, and the least contribution to global warming was made by methanol obtained from biomass. LNG is superior to other fuels with zero $PM_{2.5}$ production. On the other hand, the total energy consumption of all alternative fuels is higher than HFO, and methanol obtained from GTL and natural gas contributes to global warming more than HFO (Brynolf 2014).

In a study in which the environmental effects of LNG and methanol were evaluated from a life cycle perspective, it was determined that both fuels are much more environmentally friendly than HFO, but they do not cause a serious reduction effect on climate change (Brynolf et al. 2014).

In a study examining the life cycle performance of LNG, it was stated that LNG provides better results in short-distance transportation due to technological constraints, it does not seem very suitable for long-term use due to infrastructure constraints (supply and distribution), and it is not expected to exhibit a climate-neutral performance before 190 years (Thomson et al. 2015).

In a study in which an 1805 TEU container ship and a passenger-cargo ship capable of carrying 683 passengers and 256 TEUs sailing in the Taiwan Strait were evaluated, it was concluded that the general environmental performance of LNG use is good, but it could not give the expected effect on global warming due to methane leakage. Most of the NO_x, carbon monoxide (CO) and GHGs are generated during operation, while most of the sulfur dioxide (SO_2) and PM gases are generated during the production phase (Hua et al. 2017).

In a study in which the environmental effects of a general cargo and a tanker operating with hydrogen and ammonia are examined in life cycle perspective, the use of HFO provides the worst results in all conditions, while the use of hydrogen provides better results in the titles of ecotoxicity, GWP, acidification, abiotic depletion, and ozone layer depletion is higher than ammonia (Bicer and Dincer 2018a).

In a similar study using the same ships, hydrogen again presented very good environmental performance (Bicer and Dincer 2018b). In a study examining the environmental impacts of HFO, MGO, and alternative fuels (LNG, hydrogen, methanol and vegetable oil, biodiesel, and bio-LNG from soy and rapeseed) from a life cycle perspective, although the emissions at the production stages are slightly different, HFO, MDO, and LNG have similar cumulative amounts of CO_2eq per kWh. Similarly, it was observed that methanol produced a high amount of CO_2eq per kWh during the operation phase. Thus, methanol has reached a higher emission value than fossil fuels. Although the entire production of CO_2eq was in the production phase, the worst performance was presented by liquefied hydrogen. Methanol increased GHG production by 12–15%, while biofuels and bio-LNG decreased by 57–79% and bio-LNG by 40–41%, respectively. HFO has arguably the highest emissions of PM and SO_x. Hydrogen does not produce SO_x and PM during operation but falls behind biofuels in a life cycle perspective due to emissions generated during production. Methanol did not produce SO_x and PM in any process. Biofuels have reached higher values than even HFO in NO_x production. As a result, an alternative fuel that simultaneously reduces both local emissions and GHGs is not yet available (Gilbert et al.

2018). In a study on marine biofuels, it was concluded that the region where the raw material of the biofuel is produced has a significant weight on the total environmental impact of the fuel due to various reasons (Kesieme et al. 2019).

In a recent study, in which a fuel with 1×10^7 energy capacity is used as a functional unit, LNG and MGO were evaluated from a life cycle perspective, and it was concluded that LNG produces 5–7 times less local emissions compared to MGO, and although the methane leakage puts LNG in doubt, it was also observed that the effect of the leakage on global warming remains at a negligible level (Hwang et al. 2019).

In a study considering only the production and distribution processes of LNG, methanol, DME, liquid hydrogen, and liquid ammonia, it was concluded that methanol and DME perform more environmentally friendly than LNG, while LNG seems more environmentally friendly than both in a life cycle perspective. Hydrogen, produced with electricity from renewable sources, is a very powerful alternative. Liquefied ammonia from natural gas also outperforms LNG, DME, and methanol from a life cycle perspective (Al-Breiki and Bicer 2021).

In a study examining the environmental impacts of the use of MGO, LNG, and hydrogen on a 12,000 gross tonnage (GT) ferry from a life cycle perspective, it was calculated that production and distribution of MGO produced 1.7 times more CO_2eq than LNG, while production and distribution of hydrogen led to very high GHG production. Moreover, from whatever source it is produced, hydrogen produces far more GHGs than MGO and LNG. Similarly, hydrogen obtained from coal is ahead of other fuels in acidification, eutrophication, and PM formation. Hydrogen from nuclear, electricity, and renewable sources produces fewer GHGs than MGO but higher than LNG. In the operational phase, hydrogen produces zero GHGs while LNG does higher than MGO. There is a similar situation in acidification, eutrophication, photochemical ozone formation and PM formation, and the effect of hydrogen is zero. The worst overall effect was produced by hydrogen from coal, followed by hydrogen from electricity. The best results on GHGs were performed by hydrogen from nuclear fuel and renewable energy, respectively. Accordingly, while hydrogen meets Tier III targets for NO_x, 2020 targets for SO_x, and 2050 targets for GHGs, LNG is far from 2050 targets, and MGO is far from both Emission Control Area (ECA) Tier III and 2050 targets (Hwang et al. 2020).

In another study, in which the supply chain and operation process are examined from the life cycle perspective on dual-fuel engines using traditional and alternative fuels, the use of LNG instead of HFO can increase or decrease upstream energy use by up to 2% and 3%, respectively. On the other hand, a decrease of 1–21% is observed in downstream energy use. Although the use of LNG raises doubts due to the decrease in cargo capacity, continuing use of MDO in auxiliary engines, and methane leakage, it increases energy efficiency if cargo capacity is kept constant, especially for tankers and cruise ships (Seithe et al. 2020).

In a recent study, in which all existing alternative fuels and their effects on human health as well as the environment are evaluated from a life cycle perspective, it was observed that the most impact on human health in the short, medium, and long term is presented by LNG, biodiesel, and again LNG, respectively. From the life cycle perspective, ammonia produces the most CO_2, while biogas produces the least. When the results are evaluated cumulatively, biogas is in the first place in environmental terms. It is followed by liquefied petroleum gas (LPG) and ethanol from wood. The worst fuels are methanol, biodiesel, and ammonia. In the study, it was emphasized that biogas provides a two-way gain if it is obtained from organic wastes (Bilgili 2021).

According to the results of a study examining the use of natural gas as ship fuel from a life-cycle perspective, the use of natural gas produces 2% fewer CO_2 emissions in the diesel cycle. In total emissions, although methane gas is released much more than fossil fuels due to leakage, a significant decrease in GHG emissions has been observed (Manouchehrinia et al. 2020).

A recent study includes the examination of the HyMethShip concept, which is bunkered with methanol and aims to stop carbon production by decomposing methanol into H_2 and CO_2 before it is burned. It was observed that this prototype ship significantly reduces acidification, GWP, marine eutrophication, PM formation, photochemical ozone formation, and soil eutrophication compared to its MGO counterparts (Malmgren et al. 2021).

In a study in which the environmental assessments of construction, operation, and dismantling processes of the two sister ships, were evaluated from a life-cycle perspective. Since the ships are sisters, the construction and dismantling processes are similar. The main difference is the operation process. According to the results, while the effects of the ship using diesel fuel on human health and the environment are higher, the ship using LNG has a greater impact on climate change due to methane leakage (Cucinotta et al. 2021).

In a recent study on LNG life cycle emissions, it was concluded that the use of LNG reduces life cycle emissions by 18% (Al-Douri et al. 2021). According to the results of the study, in which the life cycle calculations of electricity, methanol, LNG, hydrogen, and ammonia use in ships operating in Croatian inland waters were realized, it was concluded that the use of electricity gives the most environmentally friendly results. Methanol has been found to be the most economically efficient alternative, but a suitable and developed infrastructure must be established to get full efficiency from methanol (Perčić et al. 2021).

According to the results of a recent study on understanding the overall performance-benefit/harm balance of LNG use, dual-fuel systems contribute less to global warming, acidification, and eutrophication. Environmental impact is directly proportional to engine power. The effects of global warming, acidification, and eutrophication during the operation are 2 and 10 times higher, respectively, compared to the production and distribution processes (Jang et al. 2021).

In another study examining the effects of LNG and methanol use on GHG production, air quality and cost from a life cycle perspective, spark-ignition, four-stroke low-pressure dual-fuel, two-stroke high-pressure dual-fuel, two-stroke low-pressure dual-fuel engines (all operating with the lean mixture), and methanol engines using HFO and MDO are compared. According to the results, the two-stroke high-pressure dual-fuel system contributes the least to global warming while methanol offers the highest.

All systems running on LNG achieved reductions in SO_x emissions between 82–89% compared to HFO. Low-pressure dual-fuel systems and spark-ignition systems reduce NO_x emissions between 83 and 93%, while NO_x emissions increase by 14% in high-pressure dual-fuel systems. All LNG systems cost 20–80% less, but methanol costs 10–140% higher. The final results show that the best cumulative performance is produced by the two-stroke high-pressure dual-fuel system, while the methanol is below the expected performance (Balcombe et al. 2021).

A study based on onboard measurements indicated that while biofuel blends can reduce life cycle CO_2 emissions by 40%, CO_2, NO_x, and SO_x emissions can be reduced by 1.24%, 3%, and ~50%, respectively, during operation (Stathatou et al. 2022). Another recent study, which covers the comparison of the conventional internal combustion engine (ICE) and H_2 ICE in terms of life cycle perspective, indicated that using H_2 can achieve a 45–72% reduction in global warming potential and abiotic depletion potential (Fernández-Ríos et al. 2022). A study discussed the use of hydrogen in the maritime sector and reviewed the hydrogen storage methods and hydrogen combustion concepts on a marine diesel engine. Besides, it was indicated that because hydrogen is highly flammable, the use of hydrogen should be designed under the Code of Safety for Ships Using Gas or Other Low-flashpoint Fuels (Inal et al. 2022).

Table 2.2 summarizes the abovementioned studies.

2.3 Discussions and Conclusions

It was seen that the studies on alternative fuels have been increasing since 2000, and recently, the focus has shifted to ammonia, hydrogen, and methanol (Ampah et al. 2021). It is stated that the predicted success of IMO's 2050 targets based on alternative fuels will come with radical technological transformation, social pressure, financial incentives, and reform in local, regional, and international regulations (Mallouppas and Yfantis 2021). In the last MEPC meeting on the decarbonization of ships, it was emphasized that many zero-carbon ships have to be delivered by 2030 to achieve the 2050 targets, but it was stated that the current technological infrastructure is not sufficient (IMO 2021). On the other hand, it was stated that liquefied hydrogen (LH_2) and ammonia can cover 79–91% of current cargo operations without causing any decrease in cargo capacity. This ratio is 93% and 98% for methanol and methane (Stolz et al. 2022).

Although alternative fuels are seen as the best environmental solution in the short, medium, and long term, these fuels can be more efficient and effective when used together with technical transformations such as innovative machine modification. In this way, a decrease in the number of many pollutants can be observed, and an increase in energy efficiency can be achieved. There are studies on the effects of some alcohol-based fuels in particular (Shamun et al. 2020; Belgiorno et al. 2019; Ianniello et al. 2021; Luca et al. 2022).

Table 2.2 Summary of the presented studies

Fuels	Results
Natural Gas/Biogas/Hydrogen	Methanol/biomethanol causes the greatest environmental impact
LNG	LNG usage causes a slight reduction in GWP
HFO/MDO/MGO/LNG Methanol/DME/F-T Diesel	F-T diesel and DME are the most environmentally friendly fuels
HFO/MGO/GTL/Biodiesels	LNG performs the best environmental impact
LNG/Methanol/HFO	LNG and methanol are far more environmentally friendly than HFO
LNG	LNG is suitable for short-distance voyages
LNG	Most of the SO_2 and PM are produced during fuel production
Hydrogen/Ammonia	Hydrogen provides much better performance in various impact categories
HFO/MGO/LNG/Hydrogen Methanol/Vegetable Oil Biodiesel/Bio-LNG	While methanol increased GHG emissions, biofuels and bio-LNG provide a reduction in GHGs
LNG/MGO	LNG produces 5–7 times fewer emissions
LNG/Methanol/DME Hydrogen/Ammonia	LNG performs the best performance
MGO/LNG/Hydrogen	Hydrogen performs the best performance
LNG/HFO	LNG can reduce downstream emissions by up to 21%
Alternative Fuels	Biogas presents the best performance in terms of human health
Hydrogen/MGO	Hydrogen reduces various environmental impacts
Diesel/LNG	While LNG reduces the impacts on human health, it has a greater impact on climate change

In light of the above information and the data obtained from the reviewed papers, it is difficult to say that alternative marine fuels are currently preferable to conventional fuels. The most important factor is the technological infrastructure, which MEPC also draws attention to in its last meeting. Fuel production, distribution, supply, and use processes should be considered as a whole, and it should not be forgotten that the technological inadequacy in any of them negatively affects the others. All alternative fuels outperform conventional fuels in terms of operational emissions. Although additional measures to reduce emissions are needed in some cases, no problem is expected from alternative fuels in this regard. There is still a long way to go in financial issues. Improving technological infrastructures will provide access to more economical fuels, but there is still time for the existing structure to change. Many problems need to be overcome for the ships to be able to use the mentioned fuels in a real sense—that is, in international long-distance trade. Above all, accurate and precise calculation of the life cycle emissions and costs of fuels is also very important for revealing the true potential of these fuels.

The inadequate production, distribution, and supply processes of alternative fuels seriously undermine the superior properties of these fuels over conventional fuels. Thus, alternative fuels should be evaluated not only in terms of the operation process, but also as a chemical in a holistic way. During the evaluation process, all views submitted by the four main stakeholders in fuel acceptance—academia, ship owners, society, and government—must be carefully considered, and each stakeholder's different priorities should be duly considered. Besides, as indicated in a recent study (Kołakowski et al. 2021), studies in the future should focus more on legislative issues to provide completeness.

In addition, the social aspects of fuel switches are another problem that has to be considered. As the use of new fuels expands, new business lines and new risks will emerge. The setting of standards, salaries, and practices for these risks and lines of business should therefore be scrutinized with particular care.

The use of alternative fuels generally reduces carbon emissions. However, nitrogen-based fuels such as ammonia can increase NO_x emissions, and the methane leakage caused by the transportation of fuels such as LNG may contribute more to global warming. It can be said that all alternative fuels cause a sharp reduction in SO_x and PM emissions. On the other hand, alternative fuels are an important solution because of the need for innovative solutions to the climate change problem caused by existing fossil fuels.

The most important driving force in the global adoption of alternative fuels is, above all, environmental concerns. If fossil fuels were not environmentally risky, no one would have thought to turn to these fuels, but the only exception could be economic concerns. Therefore, as long as the pressure created by human being on the environment continues, the orientation to different energy sources will be an extremely natural and effective result. From a life-cycle perspective, however, there will eventually be an economic constraint on the use of fossil fuels, which cannot be produced but can be obtained naturally from mines or oil fields, as the extraction of fossil fuels will become more and more costly. Therefore, the orientation toward alternative fuels is an inevitable result. Although costs are the most important constraint in front of this transformation today, difficult problems such as procurement, various biases, insufficient infrastructure, and low efficiency await us.

Considering the global climate change and other environmental problems that have occupied the world for a long time, today's energy problems and economic constraints, the adoption and spread of alternative fuels become a very complex problem. Therefore, it is very important to plan every step taken correctly. The selected alternative fuel should be managed at an economically acceptable level, produce satisfactory solutions to address environmental concerns, and be safe and sustainable in terms of energy supply. The holistic solution to all these problems can be provided by LCA, which is designed to bring holistic solutions to holistic problems, not with traditional approaches that focus on problems one by one. LCA is considered to be the most appropriate solution approach for the current situation, as it includes calculation systems covering economic and social impacts as well as environmental impacts. The main problem that the use of LCA can cause is the difficulties in determining energy roadmaps, which are too complex, databases that need to be constantly updated and long-term working times.

Examining the reviewed studies holistically shows that we are still in a transitional stage, which leads the stakeholders to be in quandary, regarding the use of alternatives. However, LNG is considered a good transition fuel until 2030, and it is indicated that LNG and methanol may help to reach the determined targets in a short period. Besides, hydrogen is considered the fuel with the highest potential in the medium and long term. Although it is projected that emissions from ships will be reduced until they reach the desired level in line with the UN's SDGs, current studies and estimates indicate that more efforts have to be done within the framework of stricter regulations to reach the targets.

Evaluation of the possibility of conversion of alternative fuels to conventional fuels was determined as the focus of the study. Accordingly, in light of the reviewed information, it is recommended that the LCA method should be prioritized so that the compatibility of alternative marine fuels with the SDGs can be determined more precisely.

Acknowledgements In beloved memory of my father, Metin Bilgili, who follows my all work with great passion and constantly encourages me for deeper studies, who passed away on 19th of December, 2021. No work that I have done or will I do would be possible without his vision. Thank you, father, for all you have taught me. There is still much to learn from you.

References

Al-Breiki M, Bicer Y (2021) Comparative life cycle assessment of sustainable energy carriers including production, storage, overseas transport and utilization. J Clean Prod 279:123481. https://doi.org/10.1016/j.jclepro.2020.123481

Al-Douri A, Alsuhaibani AS, Moore M, Nielsen RB, El-Baz AA, El-Halwagi MM (2021) Greenhouse gases emissions in liquified natural gas as a marine fuel: Life cycle analysis and reduction potential. Can J Chem Eng, 1–9. https://doi.org/10.1002/cjce.24268

Ampah JD, Yusuf AA, Afrane S, Jin C, Liu H (2021) Reviewing two decades of cleaner alternative marine fuels: towards IMO's decarbonization of the maritime transport sector. J Clean Prod 320:128871. https://doi.org/10.1016/j.jclepro.2021.128871

Balcombe P, Staffell I, Kerdan IG, Speirs JF, Brandon NP, Hawkes AD (2021) How can LNG-fuelled ships meet decarbonisation targets? An environmental and economic analysis. Energy 227:120462. https://doi.org/10.1016/j.energy.2021.120462

Belgiorno G, Di Blasio G, Beatrice C (2019) Advances of the natural gas/diesel RCCI concept application for light-duty engines: comprehensive analysis of the influence of the design and calibration parameters on performance and emissions BT - natural gas engines : for transportation and power generation. Presented at the (2019)

Bengtsson S, Andersson K, Fridell E (2011) A comparative life cycle assessment of marine fuels: Liquefied natural gas and three other fossil fuels. Proc Inst Mech Eng Part M J Eng Marit Environ 225:97–110. https://doi.org/10.1177/1475090211402136

Bicer Y, Dincer I (2018a) Environmental impact categories of hydrogen and ammonia driven transoceanic maritime vehicles: a comparative evaluation. Int J Hydrogen Energy 43:4583–4596. https://doi.org/10.1016/j.ijhydene.2017.07.110

Bicer Y, Dincer I (2018b) Clean fuel options with hydrogen for sea transportation: a life cycle approach. Int J Hydrogen Energy 43:1179–1193. https://doi.org/10.1016/j.ijhydene.2017.10.157

Bilgili L (2021) Comparative assessment of alternative marine fuels in life cycle perspective. Renew Sustain Energy Rev 144:110985. https://doi.org/10.1016/j.rser.2021.110985

Brynolf S, Fridell E, Andersson K (2014) Environmental assessment of marine fuels: Liquefied natural gas, liquefied biogas, methanol and bio-methanol. J Clean Prod 74:86–95. https://doi.org/10.1016/j.jclepro.2014.03.052

Brynolf, S.: Environmental Assessment of Present and Future Marine Fuels, (2014)

Cucinotta F, Raffaele M, Salmeri F, Sfravara F (2021) A comparative life cycle assessment of two sister cruise ferries with diesel and liquefied natural gas machinery systems. Appl Ocean Res 112:102705. https://doi.org/10.1016/j.apor.2021.102705

Di Luca G, Pipicelli M, Ianniello R, Belgiorno G, Di Blasio G (2022) Alcohol fuels in spark ignition engines BT—application of clean fuels in combustion engines. Presented at the (2022)

Eyring V, Köhler HW, Lauer A, Lemper B (2005) Emissions from international shipping: 2. Impact of future technologies on scenarios until 2050. J Geophys Res D Atmos 110, 183–200. https://doi.org/10.1029/2004JD005620

Faber J, Hanayama S, Zhang S, Pereda P, Comer B, Hauerhof E, van der Loeff W, Smith T, Zhang Y, Kosaka H, Adachi M, Bonello J, Galbraith C, Gong Z, Hirata K, Hummels D, Kleijn A, Lee D, Liu Y, Lucchesi A, Mao X, Muraoka E, Osipova L, Qian H, Rutherford D, la Fuente S, Yuan H, Perico C, Wu L, Sun D, Yoo D, Xing H (2020) Fourth IMO GHG study

Fernández-Ríos A, Santos G, Pinedo J, Santos E, Ruiz-Salmón I, Laso J, Lyne A, Ortiz A, Ortiz I, Irabien Á, Aldaco R, Margallo M (2022) Environmental sustainability of alternative marine propulsion technologies powered by hydrogen - a life cycle assessment approach. Sci Total Environ 820. https://doi.org/10.1016/j.scitotenv.2022.153189

Gilbert P, Walsh C, Traut M, Kesieme U, Pazouki K, Murphy A (2018) Assessment of full life-cycle air emissions of alternative shipping fuels. J Clean Prod 172:855–866. https://doi.org/10.1016/j.jclepro.2017.10.165

Hua J, Wu Y, Chen HL (2017) Alternative fuel for sustainable shipping across the Taiwan Strait. Transp Res Part D Transp Environ 52:254–276. https://doi.org/10.1016/j.trd.2017.03.015

Hwang S, Jeong B, Jung K, Kim M, Zhou P (2019) Life cycle assessment of LNG fueled vessel in domestic services. J Mar Sci Eng 7:1–25. https://doi.org/10.3390/jmse7100359

Hwang SS, Jeong B, Jung KH, Kim M, Zhou P, Gil SJ, Lee GN, Won Lee J, Park H, Jung KH, Suh SB (2020) Life cycle assessment of alternative ship fuels for coastal ferry operating in republic of Korea. J. Mar. Sci. Eng. 8:660. https://doi.org/10.3390/JMSE8090660

Ianniello R, Belgiorno G, Di Luca G, Beatrice C, Di Blasio G (2021) Ethanol in dual-fuel and blend fueling modes for advanced combustion in compression ignition engines BT - alcohol as an alternative fuel for internal combustion engines. Presented at the (2021)

IEA (2017) CO_2 emissions from fuel combustion

IEA (2020) Key world energy statistics. Paris

IMO (2018) Adoption of the initial IMO strategy on reduction of GHG emissions from ships and existing IMO activity related to reducing GHG emissions in the shipping sector

IMO (2021) MEPC 77/7/6-Reduction of GHG Emissions from Ships

IMO (2022): Sustainable Development Goals @ www.imo.org, https://www.imo.org/en/MediaCentre/HotTopics/Pages/SustainableDevelopmentGoals.aspx

Inal OB, Zincir B, Dere C (2022) Hydrogen as maritime transportation fuel: a pathway for decarbonization BT - greener and scalable E-fuels for decarbonization of transport. Presented at the (2022)

Intergovernmental Oceanographic Commission (2021) Ocean literacy within the united nations decade of ocean science for sustainable development: a framework for action

IPCC (2021) Summary for policymakers In: Climate change 2021: the physical science basis
IRENA (2021) A pathway to decarbonise the shipping sector by 2050. Abu Dhabi
Jang H, Jeong B, Zhou P, Ha S, Nam D (2021) Demystifying the lifecycle environmental benefits and harms of LNG as marine fuel. Appl Energy 292:116869. https://doi.org/10.1016/j.apenergy.2021.116869
Kesieme U, Pazouki K, Murphy A, Chrysanthou A (2019) Attributional life cycle assessment of biofuels for shipping: addressing alternative geographical locations and cultivation systems. J Environ Manag 235:96–104. https://doi.org/10.1016/j.jenvman.2019.01.036
Kołakowski P, Gil M, Wróbel K, Ho YS (2021) State of play in technology and legal framework of alternative marine fuels and renewable energy systems: a bibliometric analysis. Marit Policy Manag 00:1–25. https://doi.org/10.1080/03088839.2021.1969460
Mallouppas G, Yfantis EA (2021) Decarbonization in shipping industry: a review of research, technology development, and innovation proposals. J Mar Sci Eng 9:415. https://doi.org/10.3390/jmse9040415
Malmgren E, Brynolf S, Fridell E, Grahn M, Andersson K (2021) The environmental performance of a fossil-free ship propulsion system with onboard carbon capture-a life cycle assessment of the HyMethShip concept. Sustain. Energy Fuels. 5:2753–2770. https://doi.org/10.1039/d1se00105a
Manouchehrinia B, Dong Z, Gulliver TA (2020) Well-to-Propeller environmental assessment of natural gas as a marine transportation fuel in British Columbia. Canada Energy Rep 6:802–812. https://doi.org/10.1016/j.egyr.2020.03.016
Øberg MM (2013) Life cycle assessment of fuel choices for marine vessels. http://www.diva-portal.org/smash/get/diva2:653083/FULLTEXT01.pdf
Perčić M, Vladimir N, Fan A (2021) Techno-economic assessment of alternative marine fuels for inland shipping in Croatia. Renew Sustain Energy Rev 148:111363. https://doi.org/10.1016/j.rser.2021.111363
Seithe GJ, Bonou A, Giannapoulos D, Georgopoulou CA, Founti M (2020) Maritime transport in a life cycle perspective: how fuels, vessel types, and operational profiles influence energy demand and greenhouse gas emissions. Energies 13:2739
Serra P, Fancello G (2020) Towards the IMO's GHG goals: a critical overview of the perspectives and challenges of the main options for decarbonizing international shipping. Sustain 12:3220. https://doi.org/10.3390/su12083220
Shamun S, Belgiorno G, Di Blasio G (2020) Engine parameters assessment for alcohols fuels application in compression ignition engines BT - alternative fuels and their utilization strategies in internal combustion engines. Presented at the (2020)
Stathatou PM, Bergeron S, Fee C, Jeffrey P, Triantafyllou M, Gershenfeld N (2022) Towards decarbonization of shipping: direct emissions & life cycle impacts from a biofuel trial aboard an ocean-going dry bulk vessel. Sustain Energy Fuels. 6:1687–1697. https://doi.org/10.1039/d1se01495a
Stolz B, Held M, Georges G, Boulouchos K (2022) Techno-economic analysis of renewable fuels for ships carrying bulk cargo in Europe. Nat Energy 7:203–212. https://doi.org/10.1038/s41560-021-00957-9
Strazza C, Del Borghi A, Costamagna P, Traverso A, Santin M (2010) Comparative LCA of methanol-fuelled SOFCs as auxiliary power systems on-board ships. Appl Energy 87:1670–1678. https://doi.org/10.1016/j.apenergy.2009.10.012
Thomson H, Corbett JJ, Winebrake JJ (2015) Natural gas as a marine fuel. Energy Policy 87:153–167. https://doi.org/10.1016/j.enpol.2015.08.027
UNCTAD (2020) Review of maritime transport. New York
UNFCCC (2015) Paris agreement. Paris
United Nations (2022): Goals @ Sdgs.Un.Org, https://sdgs.un.org/goals
Wang Y, Wright LA (2021) A comparative review of alternative fuels for the maritime sector: economic, technology, and policy challenges for clean energy implementation. World 2:456–481. https://doi.org/10.3390/world2040029

Ytreberg E, Åström S, Fridell E (2021) Valuating environmental impacts from ship emissions – the marine perspective. J Environ Manage 282. https://doi.org/10.1016/j.jenvman.2021.111958

Chapter 3
Lifecycle Emissions of Fossil Fuels and Biofuels for Maritime Transportation: A Requirement Analysis

Cagatayhan Sevim and Burak Zincir

Abstract Climate change and global warming are among the most important problems that today's world is struggling with. Greenhouse gas emissions released into the atmosphere make these problems even more intractable. The leading organization of the maritime industry, the International Maritime Organization (IMO), is taking increasingly restrictive and stricter rules and regulations on the reduction of greenhouse gas emissions, as a significant amount of greenhouse gas emissions released into the atmosphere originate from commercial ships. Therefore, researchers focused on alternative marine fuels. Although there are many types of alternative marine fuels, biofuels are the most promising fuel for a smooth transition to zero-carbon alternative fuels. This is because biofuels can be burned in existing diesel-powered ships without any modifications or with minor modifications. Existing rules that seek to control emissions mainly monitor emissions from combustion of the fuel at the end user but are likely to take into account the entire lifecycle emissions of the fuel in the coming years. For this reason, in this paper, information about the stages and processes of lifecycle assessment is given. Then, the lifecycle emissions of fossil fuels, which are widely used today, and biofuels, which have an important position both in the decarbonization of maritime transportation and in the transition to zero-carbon alternative fuels, are examined. The aim of this study is to emphasize the importance of the lifecycle assessment model in the steps to be taken to reduce greenhouse gas emissions in order to overcome the problems on a global scale and then to compare fossil fuels and biofuels for the maritime industry within the scope of lifecycle emissions.

Keywords Biofuel · Fossil fuel · Lifecycle assessment · Lifecycle emission · ILUC · Maritime transportation

---

C. Sevim (✉)
Naval Architecture and Maritime Faculty, Yildiz Technical University, Istanbul, Türkiye
e-mail: sevimca@yildiz.edu.tr

B. Zincir
Maritime Faculty, Istanbul Technical University, Istanbul, Türkiye

B. Zincir et al. (eds.), *Decarbonization of Maritime Transport*, Energy, Environment, and Sustainability, https://doi.org/10.1007/978-981-99-1677-1_3

Nomenclature

CO	Carbon monoxide
CO_2	Carbon dioxide
DCS	Data Collection System
EEDI	Energy Efficiency Design Index
EFTA	European Free Trade Association
EMSA	European Maritime Safety Agency
EU	European Union
GHG	Greenhouse Gas
HFO	Heavy Fuel Oil
ILUC	Indirect Land Use Change
IMO	International Maritime Organization
ISO	International Organization for Standardization
LCA	Lifecycle Assessment
LNG	Liquefied Natural Gas
LPG	Liquefied Petroleum Gas
MDO	Marine Diesel Oil
MRV	Monitoring, Reporting, and Verification
NO_x	Nitrogen oxides
PM	Particulate Matter
SEEMP	Ship Energy Efficiency Management Plan
SO_2	Sulfur oxides
TTW	Tank-to-Wake
UCO	Used Cooking Oil
UNCTAD	United Nations Conference on Trade and Development
WTT	Well-to-Tank
WTW	Well-to-Wake

3.1 Introduction

Transportation is a sector with a high energy need. About 19% of global energy consumption belongs to the transportation sector (Inal et al. 2022). The transportation sector includes airways, railways, roadways, and maritime transportation. Maritime transportation, on the other hand, is the most efficient type of transportation and logistics thanks to the internal combustion diesel engines used (Zincir 2022). According to data from the United Nations Conference on Trade and Development (UNCTAD), the commercial marine fleet increased by 3% between January 1 and December 31, 2021. The number of ships of 100 gross tons and above has reached 99,800 at January 2021 (Review of Maritime Report 2021). Shipping, which is an important part of international trade, is one of the fastest-growing sectors in the world. With the effect of the developing global economy, the amount of products transported by maritime transportation has increased by more than 150% since the 1990s (Baldi et al. 2020). Approximately, 90% of the products transported in global trade are provided by maritime transportation (Zincir and Deniz 2021).

Although maritime transportation is cost-effective, 99% of existing ships have internal combustion engines and use petroleum-derived fuel (Rattazzi et al. 2021). The amount of fuel consumed in maritime transportation is approximately 300 million tons, which corresponds to 7% of global fuel consumption and 3% of energy demand (IMO et al. 2014). Heavy fuel oil (HFO) 72%, marine diesel oil (MDO) 26%, and liquefied natural gas (LNG) 2% constitute petroleum-derived fuels consumed by ships (Gray et al. 2021). Depending on the petroleum-derived fuel consumption, maritime transportation emits a substantial amount of greenhouse gas (GHG), and carbon dioxide constitutes the largest majority of greenhouse gases. Ships emit 0.61% of worldwide CO emissions, 9.84% of SO_x emissions, 14.74% of NO_x emissions, and 6.75% and 3.56% of PM2.5 and PM10 emissions, consequently, according to the European Environment Agency (EEA) (2021). In addition, in its fourth greenhouse gas study, International Maritime Organization (IMO) stated that 2.89% of global CO_2 emissions originate from shipping (IMO 2020).

3.1.1 IMO Actions

The IMO, which sets emissions regulations for marine transportation, tightens them more strictly every day to reduce shipboard GHG emissions. The most important greenhouse gas is CO_2, and reducing CO_2 emissions in marine engines is a challenging issue due to the carbon-containing fuels used and the increasing number of ships (Dere and Deniz 2020). It has been determined that when the maritime industry is considered as a country, it would be the sixth largest CO_2 emitter country in the world (Balcombe et al. 2018). In order to overcome this issue, IMO adopted the first greenhouse gas strategy plan in 2018 to reduce greenhouse gases and has set a goal within this framework (Joung et al. 2020). The primary objectives of the IMO's initial strategy plan are to minimize CO_2 emissions per unit of transport activity by at least 70% by 2050 compared to 2008 and decrease worldwide ship-source emissions by 50%. This is the first to fully demonstrate the maritime industry's compliance with the goals of the Paris Agreement (American Bureau of Shipping (ABS) 2020). The strategy plan contains potential short-, mid-, and long-term methods to accomplish these fundamental goals as well. Short-term measures include the years 2018 through 2023. Mid-term measures encompass 2023 and 2030, while long-term measures cover 2030 and beyond (Rutherford and Comer 2018). These candidate measures are given in Table 3.1.

The candidate measures can be implemented without any restriction. To accomplish decarbonization in maritime transportation, one or a combination of more than one candidate measure may be applied. Countries, owners, and operators of ships are in charge of the decarbonization process. In addition, there are some international project initiatives that contribute to the achievement of IMO's greenhouse

Table 3.1 Candidate measures

Short-term candidate measures	Mid-term candidate measures	Long-term candidate measures
Enhancing the current energy efficiency framework (EEDI, SEEMP)	Program for the effective adoption of zero- and low-carbon fuel alternatives	Pursue the creation and supply of fossil-free or zero-carbon fuels
The creation of operational and technological energy efficiency solutions for both new and used ships	Measures to improve operational energy efficiency for both new and old ships	Encourage and assist the widespread adoption of any other new or creative emission reduction techniques that may be available (s)
Creation of an existing fleet improvement program	Market-based actions, novel emission reduction mechanisms	
Speed optimization/reduction	Continued development and improvement of technical collaboration and capacity building	
Taking action against volatile organic compounds	Creation of a feedback mechanism to allow for the learning of lessons from the use of measures	
Creation and revision of national action plans		
Technical cooperation and capacity building should continue and be improved		
Port development and activity measures		
Launch of R&D initiatives in marine propulsion, alternative low- and zero-carbon fuels, and cutting-edge technologies		
Rewards for adopting new technologies first		
Creation of robust fuel lifecycle GHG/carbon intensity guidelines		
Conduct more research on GHG emissions		

gas strategy plan targets. These initiatives are GreenVoyage2050, Global Industry Alliance (GIA), Global Maritime Network (GMN) of Maritime Technology Cooperation Centers, Germany Asia Maritime Transport Emissions (Blue Solutions Project), GHG-SMART, NextGEN project, etc. (International Maritime Organization (IMO) 2022).

The IMO's initial GHG strategy includes operational measures, policy measures, and, more generally, technology-based measures. The use of alternative marine fuels or fuel cells is one of these strategies, but it is not the only one. Other steps include status monitoring, efficient shipboard operations, and more efficient ship designs (Dere and Deniz 2020). The technology-based measure that is prevalent at the short-, mid-, and long-term measures is the usage of alternative fuels in place of traditional fossil fuels. The strategy fosters the implementation of alternative low-carbon and zero-carbon fuels in the short-term and supports initiatives on these fuels in the mid-term. Then, it works to enhance and implement fossil-free or zero-carbon fuels in the long run. Researchers are being pushed to look for alternative marine fuels made from clean and renewable sources as a result of all these new rules, regulations, incentives, and measurements surrounding emissions.

Decarbonized marine transportation may be achieved using a variety of alternative fuels, including liquefied natural gas (LNG), liquefied petroleum gas (LPG), methanol, ethanol, hydrogen, ammonia, completely electric vehicles, and biofuels (Ryste et al. 2019). Each fuel route has pros and cons when it comes to using it in ships because of its unique characteristics. Alcohol fuels such as ethanol and methanol, among these fuels, have the potential to reduce carbon emissions by using them as fuel additives (Shamun et al. 2020), just like biofuels. Therefore, characteristic features are decisive for which alternative fuels can be used in maritime transportation in the short-, mid-, and long-term.

3.1.2 Transition to Alternative Fuels

Today, the most used marine fuels are HFO and MDO. After HFO and MDO, LNG is the third fuel of choice for marine transportation. In addition, methanol is also the fourth-most common alternative fuel consumed on vessels (Deniz and Zincir 2016). The ratio of alternative fuel vessels ordered for construction to all vessels ordered increased from 6% in May 2019 to almost 12% in June 2021. According to DNV GL, a reputable classification agency in the maritime transportation industry, there are 216 LNG-fueled vessels currently in operation, and 399 more are being ordered. 10 methanol-fueled vessels are currently in service and 29 more are being ordered, and there are six LPG-fueled vessels currently in operation and 90 more are being ordered (Hammer et al. 2050). When we consider all these, we can see that the transition to alternative fuels in the maritime sector has started and is increasing rapidly.

Although the transition to alternative fuels will be easier for new ships to be built, it will not be an easy process for existing ships. The vast majority of existing ships have engines and systems suitable for the storage and consumption of HFO and MDO, which are still widely used today (Hammer et al. 2050). Ensuring the operation of these ships with a different alternative fuel may require changes in the main engine, auxiliary engines, and other systems (in matters such as safety, storage, transfer, filter, and separation) depending on the nature of the fuel to be used. In this chapter, biofuels were examined because biofuels have a significant advantage in the transition to alternative fuels for decarbonizing maritime transportation thanks to the ability that they can be blended with petroleum-derived conventional fuels or used directly as a drop-in in the existing ship infrastructure.

3.2 Role of Biofuels in Transition Period

Engines and ships compatible with zero-carbon alternative fuels (such as hydrogen ammonia) are in the maritime sector's future plans in order for the IMO to achieve its GHG targets. However, using fuels such as hydrogen and ammonia on existing ships is not an easy task. Because major modifications need to be made in the engine and systems of the ship. These modifications are costly operations. In order to use the capital spent on currently operating ships in the most efficient way, it should be ensured that these ships operate in accordance with the new regulations and rules until the end of their life. The average life span of a ship is accepted as 30 years (Laso et al. 2018). From this point of view, most of the existing ships should be operational for many more years. According to the UNCTAD review of maritime transport 2021 report, almost fifty percent of the existing ships are 14 years old and below. Age distribution according to the types of ships is given in Table 3.2 (Review of Maritime Report 2021).

The vast majority of ships given in the table are diesel-powered ships. The old ones of these ships can be destroyed after completing their life span, but it is an important issue that the younger ones are operated in accordance with the new rules and regulations. One of the effective ways to operate these ships in accordance with the new rules is to use a drop-in transition fuel. Biofuels are alternative marine fuels that can be used as drop-in ABS (2021a). Finding biofuels that could be easily used in two-stroke or four-stroke marine diesel engines with little to no modification is one of the transitional solutions to address the emissions-reduction targets of the IMO. For this reason, an increasing number of shipping companies have started to test biofuels or their blends.

Table 3.2 Age distribution of existing ships

Ship type	Age of ship (%)				
	0–4	5–9	10–14	15–19	More than 20
Bulk carrier	18	37	24	10	10
Container	14	19	32	17	17
General cargo	5	10	16	9	59
Oil tankers	14	17	21	13	35
Others	10	17	17	9	47
All ships	11	18	19	10	41

The term "biofuel" refers to a broad range of fluids or gases that may be created from a wide range of biomass or biological wastes. For example, biofuels can be made from lignocellulosic biomass (such as miscanthus, corn stalks), used cooking oils (UCO), tallow and microalgae biomass, as well as edible vegetable oils such as palm, soybean and rapeseed (Mohd Noor et al. 2017; Lin and Lu 2021). According to the raw material utilized and the manufacturing technique, biofuels are expressed in three generations. Most of the crops in the first generation are edible food ones. Lignocellulosic raw materials and non-edible plant oils make up second-generation biofuels. The biomass of microalgae is used to produce the third generation of biofuel, and research is currently ongoing (Sevim and Zincir 2022). Biofuels can be produced from different raw materials as well as by going through different processes (Yuan-rong Zhou 2020). In Fig. 3.1, biofuels produced from different raw materials by different processes are illustrated (ABS 2021b).

Biofuels, like conventional fossil fuels, are fuels containing carbon. As can be seen in Fig. 3.1, many different biofuels are obtained according to the raw material and production process used in their production. For this reason, the emissions of the final biofuel obtained as a result of combustion are also different. In experimental studies, gains were obtained in carbon emissions as a result of combustion, depending on the mixing ratio (Abed et al. 2019; Kaya and Kökkülünk 2020). However, in experimental studies, there have been cases where there has been an increase, especially in CO_2 emission values (Xue et al. 2011). The main gain of biofuels in terms of emissions comes out with the lifecycle assessment thanks to the ability of carbon

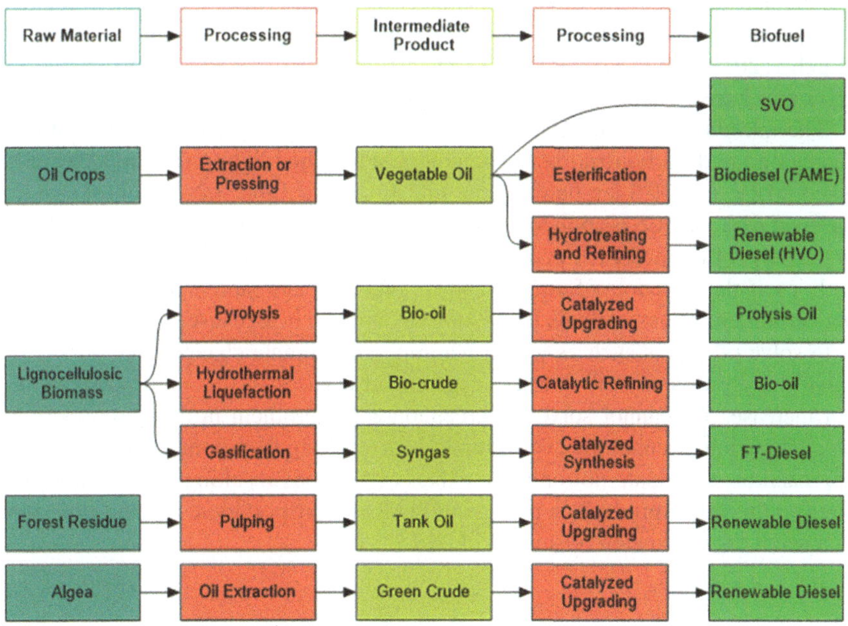

Fig. 3.1 Biofuel production pathways from different raw materials

uptake (Sevim and Zincir 2022). When we use petroleum-derived fuels, we burn fossilized hydrocarbons that are not related to the atmosphere no longer and release the carbon content into the atmosphere in gaseous form. Thus, we increase the rate of CO_2, which is the most important greenhouse gas, in the atmosphere and take action that exacerbates the problems of global warming and climate change. The situation is different for biofuels. The raw materials used to obtain biofuels are plants that are capable of biogenic carbon capture with the ability to photosynthesis. Thanks to this ability, biofuels are considered carbon neutral (Khan et al. 2021). Carbon–neutral fuels such as biofuels have an essential place in the sustainable energy transition and green economy.

3.3 Emissions Monitoring and Reporting Infrastructure

The exhaust emissions from combustion aboard ships are the main focus of the existing monitoring and regulatory infrastructure in marine transportation. The monitoring, reporting, and verification of CO_2 emissions from maritime transportation are currently governed by two regulations. The first of these is the European Union MRV Regulation (EU MRV). The MRV regulation came into effect on July 1, 2015. The European Union, along with the MRV regulations, has also led the way in producing more effective solutions to the IMO within the legal framework (Boviatsis and Tselentis 2019).

Another method is the IMO Data Collection System (IMO DCS). IMO adopted the new regulation for ships to monitor and report their fuel consumption, and on March 1, 2018, IMO DCS came into effect (Rony et al. 2019; IMO 2018). Table 3.3 gives the characteristics of MRV and DCS systems.

The introduction of the MRV and DCS systems has created an important data source on the decarbonization of shipping and has shed light on future steps to be taken. MRV and DCS are on the same line as seen in Table 3.2. The most striking difference is that DCS is applied in all international ports, while MRV is applied only in the ports of EU and EFTA countries (Zincir and Deniz 2021).

Even if there is not a formal mandate yet, it makes more sense to assess the net carbon emissions of fuels using an LCA approach. Because the main purpose is to solve problems such as global warming and climate change, and in order to solve these problems, the density of greenhouse gases in the atmosphere should not increase. We cannot solve the greenhouse gas problem in the atmosphere by only taking into account the exhaust emissions resulting from combustion. The LCA approach is to evaluate not only the exhaust emissions resulting from combustion but also the whole production process of the fuel, transportation, and the emissions as a result of combustion (Osman et al. 2021).

Table 3.3 Features of MRV and DCS systems

	IMO DCS	EU MRV
Entry into force	1 March 2018	1 July 2015
Initial monitoring period	2019	2018
Procedure	Data collection plan	Monitoring plan
Verification	Authorized organizations or flag administrations	Verifiers who are accredited independently
Target ships	Ships with a gross tonnage of 5000 or more	Ships with a gross tonnage of 5000 or more
Scope	International	EU and EFTA ports
Centralized fuel consumption database	IMO management database of fuel consumption (GISIS)	EMSA database (THETIS)
Reporting	Total fuel consumption Distance traveled Hours underway Design deadweight	Fuel consumption (port/sea) Carbon emissions Actual cargo carried Voyage distance Sea time not include anchorages

3.4 Lifecycle Assessment in General

The concept of lifecycle assessment emerged in the late 1980s from competition among manufacturers attempting to persuade users about the superiority of one product choice over another. It became clear that several techniques were being adopted in relation to the essential components in the LCA analysis as additional comparative studies with divergent conclusions were published. These components are boundary conditions, data source, and definition of the functional unit. The International Standards Organization (ISO) has prepared a set of LCA standards under the environmental management series to standardize these approaches and techniques (By Reilly-Roe and A Ltd 2012). This standard is ISO 14040, and it has also developed as the 14041-14043 standards in 1997–2000, which also states its requirements for each phase. These phases are as follows:

i. Goal and scope definition.
ii. Inventory analysis.
iii. Impact assessment.
iv. Interpretation.

These standards, valid until 2006, were re-evaluated and revised, and updated for LCA principles (ISO 14040:2006), for requirements (ISO 14044:2006). In addition, ISO has published guidelines (ISO 14047-14049) to assist in the good implementation of the LCA (Tanzer et al. 2019).

By using the LCA model for fuels, the lifecycle emissions of fuel can be estimated. The LCA approach used to determine the net carbon emissions of fuels is analyzed

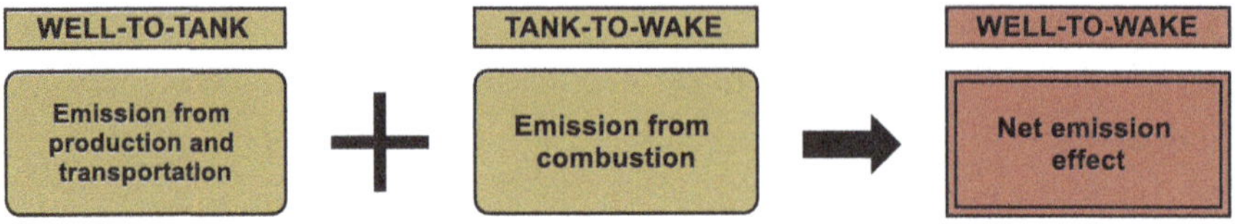

Fig. 3.2 Lifecycle emission of fuel

in 2 parts. Emissions caused by combustion on the end user are known as tank-to-wake (TTW) emissions, while emissions from fuel generated from scratch up until it reaches the ultimate consumer are known as well-to-tank (WTT) emissions. The value obtained by summing WTT and TTW emissions is the net carbon emission effect of the fuel and is called well-to-wake (WTW) (ABS 2021a). It is visually schematized in Fig. 3.2.

There are different forms of these nomenclatures in the literature. For example, cradle to grave for WTW, well to pump for WTT (Carneiro et al. 2017). The important point is not the nomenclature, but knowing the fact that emissions from production and transportation make up the 1st part, emissions from combustion or using it as an energy source make up the 2nd part, and their sum is the net carbon emissions of fuel.

3.4.1 LCA Approach for Fossil Fuels

Fossil marine fuels have been used predominantly in the maritime industry for many years. The maritime industry switched from coal to marine diesel oil in the 1920s and then to HFO in the 1950s (Balcombe et al. 2018). However, at the point that we have reached, the tightening emission rules to reduce carbon intensity have focused on emissions from these fuels. The LCA approach is generally applied to alternative marine fuels. Since HFO and MDO are reference fuels for shipping, the LCA approach applied to these fuels is important because it is comparable to that of alternative fuels (Bengtsson et al. 2008). Figure 3.3 illustrates the LCA approach to a petroleum-derived fuel.

The first stage begins with the extraction of crude oil, which is the raw material. It continues with the transportation of the extracted crude oil to the oil refinery and the production of products such as heavy fuel oil and diesel oil for the marine industry and kerosene, gasoline, and LPG for the other transportation sectors. Afterward, the process of storing and transporting the obtained fuel has completed the WTT phase with its delivery to the end users. There is energy consumption in all of these stages. Electricity is mostly used in oil extraction and refining processes. Even the source of the electrical energy used and how it is produced affect net carbon emissions. The electricity produced on a regional and country basis can be renewable energy (wind, solar, etc.), as well as nuclear, coal power plants, or natural gas conversion power plants. As a result, it is important to know the amount and proportionally the source

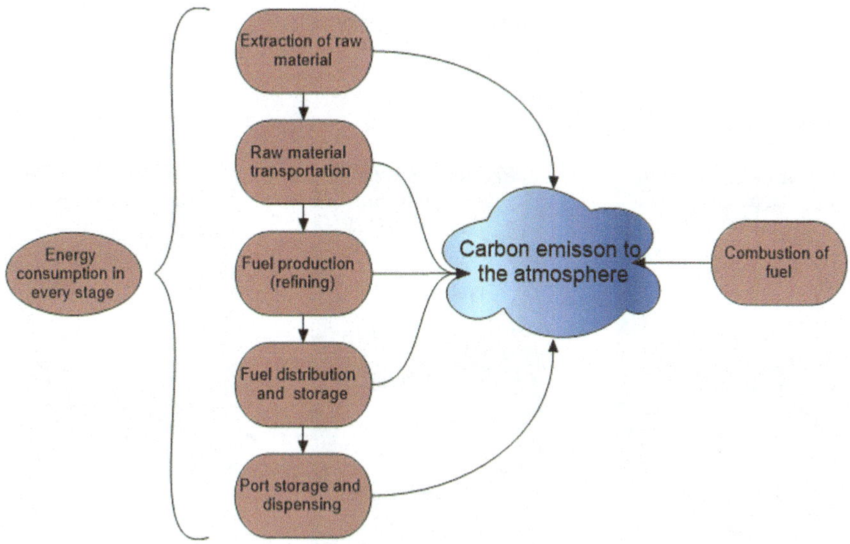

Fig. 3.3 Life cycle of conventional fossil fuels

of the electricity used in these stages. Diesel oil is mostly used as an energy source during the transportation, storage, and distribution stages of the fuel. For this reason, the emissions of the fuel used in these stages should also be taken into account when trying to determine the net carbon impact of marine fuel.

3.4.2 LCA Approach for Biofuels

Although the LCA approach in biofuels is similar to that of fossil fuels in general terms, it contains some differences (Keller et al. 2022; McKone et al. 2011). The raw material of fossil fuels can be directly extracted from oil wells. However, the raw materials to be used in biofuels require cultivation (Liu et al. 2018). The LCA process for biofuels essentially has three key phases. Phase 1 involves the cultivation of biomass, phase 2 involves the manufacturing of biofuel, and phase 3 involves the consumption of biofuel. Each phase needs energy, and the majority of this energy comes from diesel oil. For example, tractors and agricultural equipment to grow and harvest biomass, heavy-duty vehicles used to transport raw materials, intermediates and produced fuel, etc. Each step results in the emission of carbon into the environment.

Figure 3.4 illustrates the LCA approach for biofuels. As can be seen from Fig. 3.4, biomass cultivation is the 1st stage and includes the harvesting of the crops planted using agricultural land with agricultural machinery and transportation to the biofuel production facility after harvesting. The second stage is the production stage of

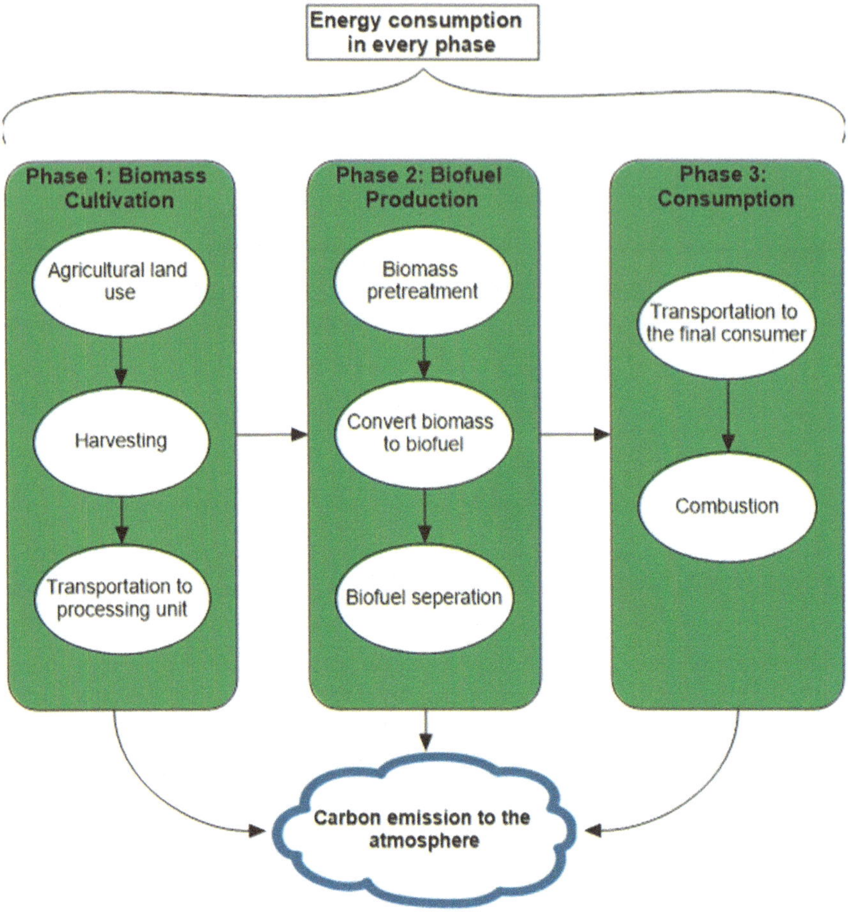

Fig. 3.4 Life cycle of biofuels

biofuel. The biomass arriving at the production facility is pre-processed to obtain an intermediate product. This intermediate product is then converted into biofuel. You can observe in detail the raw materials, pre-processes, intermediate products, and obtained biofuels in Fig. 3.1 above. The third stage includes the delivery and combustion of the biofuel to the end user.

When determining the net carbon effect of biofuels with the LCA approach, the biogenic carbon capture ability of the biomass used in the production of raw materials should also be taken into account. Plants perform photosynthesis by using the free CO_2 in the atmosphere, thanks to their unique ability. In this way, they contribute positively to the net carbon emission of the biofuel obtained from the biomass of these plants, by reducing the concentration of CO_2, one of the most important greenhouse gases in the atmosphere.

The indirect land-use change (ILUC) should also be taken into account when assessing the lifecycle assessment of biofuels. When cropland that was once utilized for food and feed is exploited to gather biomass, ILUC results (Sevim and Zincir 2022). New agricultural areas will be required to satisfy this need since the human and animal populations will continue to grow. Wetlands and forests will be turned into agricultural land due to the demand for more agricultural land (Juncker 2019; Gas et al. 2015). However, because of their capacity to absorb biogenic carbon through photosynthesis, forests play a significant role in the natural carbon cycle. Because of this, it is undesirable to convert forests to agricultural lands, and the ILUC has a negative effect on a fuel LCA (Finkbeiner 2014).

3.4.3 LCA Approach Comparison of Biofuels and Fossil Fuels

Although fossil fuels and biofuels have basically the same lines in terms of LCA approach, there are points where biofuels differ due to their unique properties. Table 3.4 gives a comparison of fossil fuels and biofuels in terms of LCA approach. While the red color has a negative meaning on the net carbon emission effect of the fuel, the green color indicates that it contributes positively. The stages of energy use are separated as WTT and TTW, and other effects are also indicated.

When we examine WTT emissions according to the table, it is similar to evaluating fossil fuels and biofuels in the storage, distribution, and transportation stages. However, the raw material procurement and production stages are not similar. While the raw material of fossil fuels is extracted from the oil well, processes such as growing and harvesting the plant to be used for the raw material of biofuels are required.

On the other hand, the production process differs according to the raw materials used in biofuels and the final product to be obtained. For this reason, varying amounts

Table 3.4 Comparison of LCA approach

	Conventional fossil fuels	Biofuels
Well-to-tank emissions	Extraction raw materials	Cultivation raw materials
	Production (refining)	Production process (differs in the type of biofuel)
	Fuel storage and distribution	Biofuel storage and distribution
	Transportation	Transportation
Tank-to-wake emission	Combustion of fuel	Combustion of biofuel
Other impacts	None	Biogenic carbon uptake
	None	ILUC
Total GHG emissions (WTW)	Sum of the above	Sum of the above

of energy consumption and emissions occur depending on which raw material and production pathway are used. In fossil fuels, the refining process of crude oil is more mature and stable. Thus, the energy and emissions consumed in the production processes of fossil fuels can be determined more easily. TTW emission resulting from the combustion of the fuel in the engine is an important point for both fuels. It varies according to the emission performance of the fuels as a result of combustion.

It is sufficient to calculate WTT and TTW emissions when determining the net carbon impact of fossil fuels. For biofuels, there are other points that need to be taken into account. First of all, since the raw materials used to obtain biofuels are plants capable of photosynthesis, it captures the free CO_2 in the atmosphere until harvest time and incorporates it into its own structure. This unique carbon capture ability of plants has a positive impact on the net carbon emissions of biofuels under the LCA approach.

But on the other hand, there is the ILUC effect, which affects negatively. The ILUC effect is related to the raw material used in the production of biofuels. When plant oils that can be used as food are used as raw materials, the ILUC effect is high. These biofuels are 1st generation. However, the ILUC effect of waste oils used in the food industry is zero. Because there is no direct use of agricultural land for fuel production. In addition, agricultural wastes such as corn stover have no ILUC effect. Also, energy crops such as miscanthus, which can easily grow in steep lands, grasslands, and pastures, can even show a negative ILUC effect which is positive for the LCA of biofuel (Pavlenko and Searle 2018). The total GHG emission impact for biofuels is estimated by calculating the total value, including WTT and TTW emissions, as well as other impacts.

3.5 Summary

This chapter focuses on the importance of the fuel lifecycle assessment approach for the decarbonization of the maritime industry. First of all, the current situation in maritime transport has been revealed with the announced fuel consumption, the number of ships, and the GHG emission data released. The actions taken by IMO to decarbonize maritime transportation and the initial strategy plan of IMO are mentioned. Then, the transition process to alternative fuels and the importance of biofuels in this process are highlighted. Afterward, a comparative analysis of current emission monitoring and reporting regulations was included, and the LCA approach was explained for the main target. Finally, the application of the LCA approach on fossil fuels and biofuels has been evaluated with a comparative table. This chapter's main outcomes are as follows:

- Maritime, which is an important part of international trade, is growing rapidly. The increasing number of ships as a result of growth leads to an increase in the amount of fuel consumed and greenhouse gases released into the atmosphere.

- The IMO adopted its initial strategy plan in 2018, outlined its main objectives, and announced short-, mid-, and long-term candidate measures. Alternative marine fuels are expressed in all time periods of the candidate measures. Day-by-day IMO adopts stricter rules and regulations to reduce greenhouse gas emissions from shipping.
- The transition period to alternative fuels has begun in maritime transport. The most preferred alternative marine fuels to conventional marine fuels are LNG and methanol, respectively. The number of LNG and methanol-fueled ships ordered for construction is also increasing.
- Although the transition to alternative fuels will be easier for newly built ships, the transition to existing ships, which have engines and systems compatible with conventional marine fuels, will not be so easy. Because, depending on the characteristics of the alternative fuel, major and costly modifications may be required.
- Biofuels, thanks to their diesel oil-like characteristics, can be used pure or blended with conventional fuels in engines, without making any changes to the ship's engines and systems, or by making minor modifications. Thanks to this feature, biofuels have the potential to play an important role in the transition period of existing ships in maritime transportation to alternative fuels.
- Considering the age distribution of the existing ships, assuming that the life of an average ship is 30 years, approximately half of them can be operated for another 15 years. Leveraging the unique role of biofuels in the transition period to alternative fuels is essential for the efficient use of capital spent on these ships.
- Exhaust emissions from combustion on ships are the main focus of the current monitoring and reporting infrastructure. Today, there are two regulations that provide this, namely EU MRV and IMO DCS, and there is no significant difference between them.
- Although there is no official directive yet, it will be much more solution-oriented to evaluate the net carbon emissions of fuels with the LCA approach for the solution of problems such as global warming and climate change, which is the main goal. Because the LCA approach aims to determine not only emissions as a result of combustion but also the entire life cycle of the fuel and the net carbon effect on the atmosphere.
- The LCA approach adopts the estimation of total net GHG emissions, taking into account all emissions from the process from scratch to fuel consumption at the end user. WTT stands for emissions from production and transportation, and TTW stands for emissions from combustion. When other effects are added to these values, the total GHG emission effect is found.
- The LCA approach to biofuels is slightly different from fossil fuels. Since the raw materials to be used to obtain biofuel are plants capable of photosynthesis, they capture the free CO_2 in the atmosphere until the harvest time and add it to their structures. This unique biogenic carbon capture ability contributes positively to the total GHG emissions of biofuels and is crucial in achieving a carbon–neutral balance.

- Another difference is the ILUC effect, which is negative. It occurs with the use of agricultural land that was previously used for food to obtain biomass. However, advanced biofuels contain raw materials that do not have an ILUC effect. Turning to raw materials without ILUC effect can be a problem solver at this point.

Acknowledgements This work was supported by the Research Fund of Istanbul Technical University (Project ID: 43655). The authors declare that they have no known competing financial interests.

References

Abed KA, Gad MS, el Morsi AK, Sayed MM, Elyazeed SA (2019) Effect of biodiesel fuels on diesel engine emissions. Egypt J Pet 28(2):183–188. https://doi.org/10.1016/j.ejpe.2019.03.001

ABS (2021a) Sustainability whitepaper—biofuels as marine fuel (May), 28

ABS (2021b) Low carbon shipping setting the course to pathways to sustainable shipping

American Bureau of Shipping (ABS) (2020) Setting the course to low carbon shipping - pathways to sustainable shipping, p 104

Balcombe P et al (2019) How to decarbonise international shipping: options for fuels, technologies and policies. Energy Convers Manag, vol 182, no December 2018, pp 72–88. https://doi.org/10.1016/j.enconman.2018.12.080

Baldi F, Moret S, Tammi K, Maréchal F (2020) The role of solid oxide fuel cells in future ship energy systems. Energy 194. https://doi.org/10.1016/j.energy.2019.116811

Bengtsson S, Andersson K, Fridell E (2008) Life cycle assessment of marine fuels. A comparative study of four fossil fuels for marine propulsion

Boviatsis M, Tselentis B (2019) A comparative analysis between EU MRV and IMO DCS—the need to adopt a harmonised regulatory system. In: 16th international conference on environmental science and technology, no. September, pp 2018–2019

By Reilly-Roe P, A. Ltd (2012) Life cycle analysis of transportation fuel pathways IEA advanced motor fuels annex 40

Carneiro MLNM et al (2017) Potential of biofuels from algae: comparison with fossil fuels, ethanol and biodiesel in Europe and Brazil through life cycle assessment (LCA). Renew Sustain Energy Rev 73(January):632–653. https://doi.org/10.1016/j.rser.2017.01.152

Deniz C, Zincir B (2016) Environmental and economical assessment of alternative marine fuels. J Clean Prod 113(X):438–449. https://doi.org/10.1016/j.jclepro.2015.11.089

Dere C, Deniz C (2020) Effect analysis on energy efficiency enhancement of controlled cylinder liner temperatures in marine diesel engines with model based approach. Energy Convers Manag 220:113015. https://doi.org/10.1016/j.enconman.2020.113015

European Environment Agency (EEA) (2021) Emissions of air pollutants from transport https://www.eea.europa.eu/data-and-maps/indicators/transport-emissions-of-air-pollutants-8/transport-emissions-of-air-pollutants-8

Finkbeiner M (2014) Indirect land use change—help beyond the hype? Biomass Bioenergy 62:218–221. https://doi.org/10.1016/j.biombioe.2014.01.024

Gas G, LC Assessment, Change DL, Bhatia SC (2015) Indirect land-use change issues relating to biofuels

Gray N, McDonagh S, O'Shea R, Smyth B, Murphy JD (2021) Decarbonising ships, planes and trucks: an analysis of suitable low-carbon fuels for the maritime, aviation and haulage sectors. Adv Appl Energy 1:100008. https://doi.org/10.1016/j.adapen.2021.100008

Hammer LS et al (2021) Maritime forecast 2050, energy transition outlook

IMO (2020) Fourth IMO greenhouse gas study: executive summary. In: IMO greenhouse gas study, vol 4, no 1, p 46
IMO Data Collection System Overview (2018). https://www.verifaviashipping.com/uploads/files/VERIFAVIA%20SHIPPING%20_%20IMO%20DCS.pdf. Accessed 29 June 2022
IMO et al (2014) The third IMO greenhouse gas study 2014. International Maritime Organization (IMO), p 327. https://doi.org/10.1007/s10584-013-0912-3.
Inal OB, Zincir B, Deniz C (2022) Investigation on the decarbonization of shipping: an approach to hydrogen and ammonia. Int J Hydrogen Energy 47(45):19888–19900. https://doi.org/10.1016/j.ijhydene.2022.01.189
International Maritime Organization (IMO) (2022) Cutting GHG emissions from shipping - 10 years of mandatory rules https://www.imo.org/en/MediaCentre/PressBriefings/pages/DecadeOfGHGAction.aspx
Joung T-H, Kang S-G, Lee J-K, Ahn J (2020) The IMO initial strategy for reducing greenhouse gas (GHG) emissions, and its follow-up actions towards 2050. J Int Marit Saf, Environ Aff, Shipp 4(1):1–7. https://doi.org/10.1080/25725084.2019.1707938
Juncker J-C (2019) ILUC. In European Commission Delegated Regulation C, 2055 final, pp 2013–2015
Kaya C, Kökkülünk G (2020) Biodiesel as alternative additive fuel for diesel engines: An experimental and theoretical investigation on emissions and performance characteristics. Energy Sources, Part a: Recovery, Utilization and Environmental Effects 00(00):1–23. https://doi.org/10.1080/15567036.2020.1774685
Keller H, Fehrenbach H, Rettenmaier N, Hemmen M (2022) Extending LCA methodology for assessing liquid biofuels by phosphate resource depletion and attributional land Use/land use change. In: Klos, Z.S., Kalkowska, J., Kasprzak, J. (eds) Towards a sustainable future – life cycle management. Springer, Cham. https://doi.org/10.1007/978-3-030-77127-0_11.
Khan N, Sudhakar K, Mamat R (2021) Role of biofuels in energy transition, green economy and carbon neutrality. Sustainability 13(22). https://doi.org/10.3390/su132212374
Laso J, Vázquez-Rowe I, Margallo M, Crujeiras RM, Irabien Á, Aldaco R (2018) Life cycle assessment of *European anchovy* (*Engraulis encrasicolus*) landed by purse seine vessels in northern Spain. Int J Life Cycle Assess 23(5):1107–1125. https://doi.org/10.1007/s11367-017-1318-7
Lin C-Y, Lu C (2021) Development perspectives of promising lignocellulose feedstocks for production of advanced generation biofuels: A review. Renew Sustain Energy Rev 136:110445. https://doi.org/10.1016/j.rser.2020.110445
Liu H, Huang Y, Yuan H, Yin X, Wu C (2018) Life cycle assessment of biofuels in China: status and challenges. Renew Sustain Energy Rev 97(March):301–322. https://doi.org/10.1016/j.rser.2018.08.052
McKone TE et al (2011) Grand challenges for life-cycle assessment of biofuels. Environ Sci Technol 45(5):1751–1756. https://doi.org/10.1021/es103579c
Mohd Noor CW, Noor MM, Mamat R (2018) Biodiesel as alternative fuel for marine diesel engine applications: a review. Renew Sustain Energy Rev 94(February 2017):127–142. https://doi.org/10.1016/j.rser.2018.05.031
Osman AI, Mehta N, Elgarahy AM, Al-Hinai A, Al-Muhtaseb AH, Rooney DW (2021) Conversion of biomass to biofuels and life cycle assessment: a review, vol. 19, no. 6. Springer International Publishing. https://doi.org/10.1007/s10311-021-01273-0
Pavlenko N, Searle S (2018) A comparison of induced land- use change emissions estimates from energy crops, no. February
Rattazzi D, Rivarolo M, Massardo AF (2021) An innovative tool for the evaluation and comparison of different fuels and technologies onboard ships. In: E3S web of conferences, vol 238. https://doi.org/10.1051/e3sconf/202123808001
Rony AH, Kitada M, Dalaklis D, Ölçer AI, Ballini F (2019) Exploring the new policy framework of environmental performance management for shipping: a pilot study. WMU J Marit Aff 18(1):1–24. https://doi.org/10.1007/s13437-019-00165-z

Rutherford D, Comer B (2018) International maritime organization's initial greenhouse gas strategy. In: International council on clean transportation (ICCT), no. April 2018, p. 8, 2018, https://theicct.org/publications/IMO-initial-GHG-strategy
Ryste JA, Wold M, Sverud T (2019) Comparison of alternative marine fuels, pp 1–65
Sevim C, Zincir B (2022) Biodiesel and renewable diesel as a drop-in fuel for decarbonized maritime transportation. In: Agarwal AK, Valera H (eds) Potential and challenges of low carbon fuels for sustainable transport. Springer, Singapore, pp 319–345. https://doi.org/10.1007/978-981-16-8414-2_10
Shamun S, Belgiorno G, Di Blasio G (2020) Engine parameters assessment for alcohols fuels application in compression ignition engines. In Alternative fuels and their utilization strategies in internal combustion engines, Y. C. and M. N. N. and A. A. K. Singh Akhilendra Pratap and Sharma, Ed. Singapore: Springer Singapore, 2020, pp. 125–139. https://doi.org/10.1007/978-981-15-0418-1_8.
Tanzer SE, Posada J, Geraedts S, Ramírez A (2019) Lignocellulosic marine biofuel: Technoeconomic and environmental assessment for production in Brazil and Sweden. J Clean Prod 239. https://doi.org/10.1016/j.jclepro.2019.117845
United Nations Conference on Trade and Development (UNCTAD) (2021) Review of maritime report 2021. http://unctad.org/en/PublicationsLibrary/rmt2015_en.pdf
Xue J, Grift TE, Hansen AC (2011) Effect of biodiesel on engine performances and emissions. Renew Sustain Energy Rev 15(2):1098–1116. https://doi.org/10.1016/j.rser.2010.11.016
Yuanrong Zhou NP (2020) The potential of liquid biofuels in reducing ship emissions|International Council on Clean Transportation. Int Council Clean Transp (September, 2020). https://theicct.org/publications/marine-biofuels-sept2020
Zincir B (2022) Ammonia for decarbonized maritime transportation. In Clean fuels for mobility, A. K. and B. G. and S. P. C. di Blasio Gabriele and Agarwal, Ed. Singapore: Springer Singapore, 2022, pp. 171–199. https://doi.org/10.1007/978-981-16-8747-1_9.
Zincir B, Deniz C (2021) Methanol as a fuel for marine diesel engines. In: Alcohol as an alternative fuel for internal combustion engines, Shukla PC, Belgiorno G, di Blasio G, Agarwal AK (eds). Singapore: Springer Singapore, 2021, pp. 45–85. https://doi.org/10.1007/978-981-16-0931-2_4

Chapter 4
Hydrogen Fueled Engine Technology, Adaptation, and Application for Marine Engines

Caglar Dere

Abstract In the last few years, there is a noteworthy increase in the number of ships that use alternative fuels apart from diesel as LNG, methanol, etc., to meet the carbon emission mitigation strategies. Hydrogen (H_2) is a contemporary promising fuel solution to cope with strict carbon emission limits. The adaptation of conventional to alternative fuels with a familiar technology, the internal combustion engine (ICE) technology, is a reasonable solution for near future carbon strategies because of being a relatively prevailing technology compared to other hydrogen sourced power generators. This chapter analyzes the potential transition of the H_2 fuel engines according to the combustion capability of hydrogen in marine engines. The lack of knowledge of hydrogen combustion and ignition in the reciprocating engines is still a main challenge of the hydrogen engines. Hydrogen combustion properties and induction into ICEs are evaluated considering advantages and challenges. The applications show that LNG-powered engines significantly have similarities with hydrogen engines and enable the transition of H_2 combustion in the marine engines. On the other hand, hydrogen engines suffer from low volumetric efficiencies and pre-ignition problems but they are capable of operate from ultra-lean to ultra-rich combustions. Thanks to the turbochargers, the stoichiometry levels are obtainable for engines to satisfy the high-power. Adding the hydrogen into the cylinder with liquefied/gas phase is the issue to be determined with ease of vaporization and the mixing. The combustion comes with near-zero CO_2 and soot advantages. However, neat H_2 combustion in marine engines is still far from the final product.

Keywords Decarbonization in maritime · Hydrogen injection · Hydrogen combustion · Hydrogen engines applications · Marine engines

---

C. Dere (✉)
Faculty of Naval Architecture and Maritime. Balatcik Mah, Izmir Katip Celebi University, Havaalanı Sosesi Cad. No:33/2 35620, Izmir, Turkey
e-mail: caglar.dere@ikc.edu.tr

B. Zincir et al. (eds.), *Decarbonization of Maritime Transport*, Energy, Environment, and Sustainability, https://doi.org/10.1007/978-981-99-1677-1_4

Abbreviations

BDC	Bottom Dead Center
CNG	Compressed Natural Gas
EGR	Exhaust Gas Recirculation
GHG	Greenhouse Gas
HFO	Heavy Fuel Oil
ICE	Internal Combustion Engine
IMO	International Maritime Organization
LNG	Liquefied Natural Gas
RoRo	Roll-on/Roll-off
TDC	Top Dead Center
NO_x	Nitrogen Oxides

4.1 Introduction

There is a fossil fuel dependency to meet the energy demand of the industry and transportation today. For an environmental concern, there is a continuous improvement need in exhaust emissions so as to reduce climate effect of power generation. The need drives the power generation sources to alternative fuels. There are many fuels with less carbon content which causes less carbon footprint on the environment, as natural gas, methanol, etc. (Zincir et al. 2019). Additionally, alcohol-based fuels, as ethanol and methanol, also have potential in use to reduce carbon emissions in ICEs (Shamun et al. 2020; Ianniello et al. 2021). However, among these alternative fuels, hydrogen is the fuel, containing zero carbon, being renewable, and having unique combustion properties that looks promising fuel for future. Hydrogen as an energy source for transportation purposes is not widely used today. However, it has a great potential as an energy source, to be stored and transported with various methods, in the following decades. Therefore, it is expected that hydrogen as a fuel can steer the future of the internal combustion engines and transportation sector, both in land and sea, because of its non-carbon content, emitting no carbon dioxide, unburned hydrocarbon and carbon monoxide emissions. It can be expected that there is a negligible rate of carbon-based emission due to the inevitable operation of lubrication oil for the safe and reliable operation of internal combustion engine. The lubrication of piston liners in order to prevent wear and provide sealing, limited amount of lubrication oil, can remain on liner walls inside the cylinder. In the combustion phase, the remained oil can burn and emit negligible amount of carbon emission.

The wide usage area of hydrogen is in industrial applications. The situation is different in transportation sector. The vehicles use hydrogen which are not propelled by internal combustion engines. The most common type of hydrogen-powered vehicles uses fuel cells, and hydrogen enters an electric contact and creates a positive charge in the system. This is a promising practice in future of the marine propulsion (Inal and Deniz 2020). However, comparing hydrogen fuel cell with hydrogen fueled engines, hydrogen internal combustion engines are advantageous on initial cost, power density, and hydrogen purity to be supplied to the engine (Tsujimura and Suzuki 2017). Today, there is a need in both in the number of hydrogen-powered generators and the large power densities. So as to satisfy the requirement in near future, internal combustion engine with hydrogen fuel looks reasonable with a more cost effective way than fuel cells.

Hydrogen as a direct fuel replaces conventional fuels and carbon-based alternative fuels and can be used as both single and assistive fuel in internal combustion engines. Alternative fuels are in use together with conventional fuels as well as hydrogen-assisted operations, carried out with other fuels. The systems are developed to decrease the carbon emissions of fossil fuels and improve their combustion efficiency of marine engines. The hydrogen enrichment in internal combustion engines is carried out with diesel engines, diesel/natural gas dual-fuel engines, and natural gas engines. Engines with spark-ignition (SI) are more prone to require less effort conversion to run on hydrogen (Verhelst et al. 2009).

This chapter will focus on hydrogen fueled internal combustion engines and its applications to marine engines. Marine engines are relatively large-scale engines compared to other internal combustion engines. Its operating parameters can differ as mean effective pressure, maximum cylinder pressures, combustion temperatures, and exhaust gas properties. When the adaptation of internal combustion engine to hydrogen-powered engine is on the issue, the problems to cope with, back fire, pre-ignition problems, knock, and rapid pressure rise as a result of rapid heat release of hydrogen because of its combustion properties. It is highly probable that these problems are encountered when adaptation of marine diesel engine is on the subject.

On the other hand, the pure hydrogen combustion in internal combustion engines suffers from reduced volumetric efficiency, lower energy amount in per volume comparing to other conventional fuels, long period of auto-ignition, and high self-ignition temperature (An et al. 2013). By the reason of the disadvantages, hydrogen studies had headed for hydrogen-doped operations rather than using it as pure. Although this phenomena prevent the reduction of carbon emissions completely, as like using LNG or alcohol-based alternative fuels, it allows use thanks to its applicability in terms of application. The prevailing practice is burning hydrogen with diesel in dual-fuel combustion mode. In this approach, hydrogen is included in the system by a carburation, injection during intake process or injection into the cylinder (Saravanan and Nagarajan 2010b).

Hydrogen production is still cost more than a conventional fuel and the whole process, and considering production of the fuel is not completely zero carbon yet. The combustion properties of hydrogen make hydrogen which is a suitable fuel for in internal combustion engines. The flame speed is 5–6 times higher than diesel and being combustible with a wide range of concentrations, which is in the range of 4% to 75% concentration by volume (Akal et al. 2020). The combustible volumetric share of hydrogen corresponds 0.1 to 7.1 in terms of equivalence ratio (Wu and Wu 2012) that allowing combustion at the lower equivalence ratios and having zero carbon

Table 4.1 Comparison of hydrogen properties with other fuels (Yip et al. 2019; Zhen et al. 2020; Kobayashi et al. 2019; Herbinet et al. 2022)

Property	Hydrogen	Methanol	Ethanol	Ammonia
Carbon content (mass%)	0%	37%	52%	0%
Density (kg/m^3)	0.0899	795	790	0.7
Lower heat value (MJ/kg)	120	20.26	27	18.6
Auto-ignition temperature (K)	858	738	698	924
Stoichiometric air–fuel ratio	34.3	6.5	9	6.06
Laminar flame speed (m/s)	2.37	0.52	0.39	0.07
Adiabatic flame temperature (K)	2382	2143	2193	1800
Minimum ignition energy in the air (mJ)	0.02	0.14	0.23	–
Quenching distance (mm)	0.6	1.85	1.65	7

	Gasoline	Methane	Diesel
Carbon content (mass%)	84%	75%	86%
Density (kg/m^3)	730–780	0.83	830
Lower heat value (MJ/kg)	44.8	50.05	42.5
Auto-ignition temperature (K)	623	813	523
Stoichiometric air–fuel ratio	17.2	17.4	14.5
Laminar flame speed (m/s)	0.37–0.43	0.38	0.37–0.43
Adiabatic flame temperature (K)	2300	2225	2300
Minimum ignition energy in the air (mJ)	0.24	0.28	0.24
Quenching distance (mm)	2	2.03	–

content, drive the hydrogen fuel as a promising solution for future IC engines. The properties of hydrogen and some other conventional and alternative fuels are given in Table 4.1.

The flammability limits for H_2 have a wide range operation capability, comparing to diesel fuel particularly. Since the diesel engines' fuel–air ratio operating limits are significantly restricted by improper combustion, smoke, and unburned hydrocarbon emissions at higher equivalence ratios as 0.7–0.9, the capability of hydrogen of sustaining the combustion under very lean conditions, and outcomes of combustion with reduced pollutant emissions that are offered under such conditions are be considered advantageous (Saravanan and Nagarajan 2008).

Increased hydrogen concentration causes rapid pressure changes in power stroke of internal combustion engines, which is seen as combustion pressure rise for each crank angle. The operating parameters of the engine must be re-set according to hydrogen combustion properties in internal combustion engine. The high-speed combustion of hydrogen causes rapid heat release in combustion phase. As injection timing must be re-arranged to adjust maximum pressure and its timing. In the recent years, by the reason of prevailing combustion characteristics of hydrogen, researchers interested in combustion and storage of hydrogen. Hydrogen has a potential to be

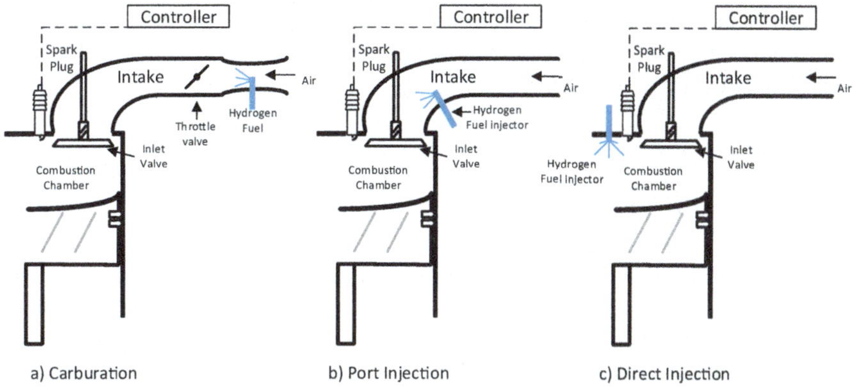

Fig. 4.1 Hydrogen injection techniques into an internal combustion engine **a** carburation method, **b** port injection, **c** direct injection

used as a single fuel in internal combustion engines, and some studies were carried out to reveal its potential about combustion performance and characteristics (Verhelst and Wallner 2009).

The hydrogen induction methods into an internal combustion engine are listed below,

- The carburation system.
- Manifold injection (timed manifold/port injection or continuous injection).
- Direct injection(high/low pressure injection).

The working principles of the injection techniques are demonstrated in Fig. 4.1, as carburation technique, port injection, and direct injection.

The systems that differ on induction system have own advantages and disadvantages. The engine power range, efficiency, safety, and emissions can be affected from the induction system characteristics. Manifold injection without timing and mixing air–fuel with carburation can be regarded as vintage technology considering hydrogen injection techniques. In the next part, the hydrogen addition methods into an internal combustion engine are introduced.

4.2 Hydrogen Induction

4.2.1 Carburation

Below this part, the common hydrogen induction technics are mentioned as carburation, port injection, and direct injection for hydrogen operations. They have their own individual advantages and disadvantages for their operations. Despite the ease of application of prior, the more flexible operation of the latter method creates self-interest for each method. Regarding the choice of the induction method, the size of the engine, required maximum power output, and the available technology implementation are determiner factors.

4.2.2 Port Injection

Port/manifold injection (timed P/M injection) systems are reasonable way to adapt a compression ignition and spark-ignited engines to neat hydrogen and hydrogen assisted operations. It does not require any major modification in the engine system. On the issue about safety, backfire and rapid heat release phenomena can be eliminated by adjusting the timed manifold injection satisfactorily (Deb et al. 2015). The injection is supplied after the exhaust valve closing (scavenging period), to ensure that, to minimize hydrogen blow up to exhaust manifold and supplied hydrogen not to mix with hot exhaust gases to prevent unintended ignition. Additionally, the air flow during scavenging provides cooling effect on the hot spots of cylinder internal parts. Low-power density, because of large unit volume of hydrogen and backfire problems, creates drawbacks for the port injection of hydrogen in to internal combustion engines. Low volumetric energy source also reduces the maximum power output of the engine because of lack of fuel/air mixture flow amount. On the other hand, the injection of low-temperature hydrogen also helps to cool inlet air. The fuel mixes with air and decreases air temperature together with unit volume. When the cryogenic hydrogen is used in the injection system, the output power level can be achievable up to gasoline engine's power.

In case of any temperature rise or unintended ignition, the hydrogen injection system is equipped temperature probe/safety controller mechanisms and flame trap, flashback arrestor to prevent backfire damages. Additionally, the system comprises pressure reducers to adjust hydrogen source high pressure. The disadvantages of hydrogen used in port injection technique create a need to study on direct injection technique so as to avoid from the disadvantages.

4.2.3 Direct Injection

Spark-ignited H_2 engines allow smooth operation even at low loads and lean conditions. The high combustion speed leads high intensity heat release within a couple of crank angles near TDC and provides higher thermal efficiency for an ICE. However, the rapid combustion together with pre-ignition phenomenon could outcome in knocking and backfire problems, at high engine loads particularly (Mohammadi et al. 2007). It can be regarded that knocking is a crucial issue for an internal combustion engine. With the injection of the fuel through ports or manifolds, there is a high potential risk of backfire and knocking. Additionally, the unit mass of hydrogen occupies more volume than any conventional fuel; for instance, 30% volume of mixture is

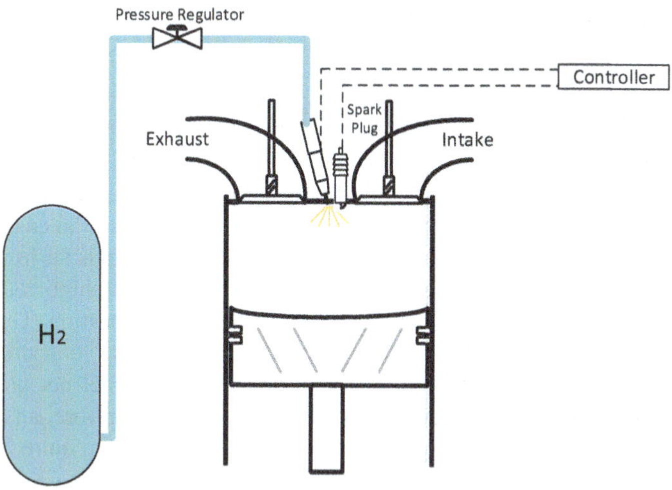

Fig. 4.2 Demonstration of spark-ignited H_2 internal combustion engine with direct injection

hydrogen when compared to 1–2% is gasoline. The problems can also result in slight decrease in the brake thermal efficiency due to the heat loss, because of early heat release-based higher combustion temperatures. In order to overcome such problems, direct injection of hydrogen rather than port fuel injection in compression stroke in to the cylinder with high-pressured system is performed by researchers (Yu et al. 2016; Kim et al. 2005).

Spark-ignited hydrogen engines with a controlled direct injection strategy and controlled ignition system is demonstrated in Fig. 4.2. In the system, during the limited period of the compression stroke, pressurized hydrogen injected in the cylinder.

The injection time, injection period, fuel amount, and pressure can be controlled by controller. The injection position and pressure must provide an adequate mixture for the ignition. The fuel injection can be during the intake period, early and late period of the compression stroke. The injection during the intake period can affect volumetric efficiency, though (Mohammadi et al. 2007). Additionally, the injection of hydrogen decreases the intake charge temperature; hence, uncontrolled ignition, namely knocking, can be less likely to occur (Boretti 2020). In this way, higher compression ratios and pressures become achievable without improper combustion. The challenge of direct injection strategies is to optimize compression ratio to perform proper combustion, injection timing so as to let enough mixing time air and hydrogen. Still the ignition time is problem which is heavily depends on load and charge conditions, injection timing can be advanced or retarded for low loads to high loads by considering operational parameters. Still, spark plug is important element in direct injection engines. Another parameter for the engines is the location of spark plug and position of injectors as location and angle (Wallner et al. 2009).

4.3 Neat Combustion of Hydrogen

Spark-Ignited H_2 Engines

Hydrogen is highly resistive to the auto-ignition. As seen in the Table 4.1, the auto-ignition temperature of hydrogen is higher than the other fuels. Therefore, operational range of pure hydrogen combustion for a compression ignition engine is very limited even in the very high compression ratios, and the problem is far from coming to a solution without any external energy source. Despite having high auto-ignition temperature, even a low ignition energy is enough to start combustion. The external energy source can be a pilot fuel "diesel fuel in common practice" or spark. Considering the pure combustion of hydrogen, a carbon contained fuel not preferred to be used in the operation to achieve zero carbon emissions. Therefore, an additional ignition source, needed as a trigger, spark plugs are used to give required energy in to cylinder contents' reaction.

Despite hydrogen combustion generating quite low emissions, with the ignition assist of pilot conventional fuel or spark-ignition together with additional advantages being near-zero carbon dioxide exhaust content, nitrogen oxide emission is the determining factor for the combustion adjustment of hydrogen in an internal combustion engine. Since the combustion temperature is higher with hydrogen, compared to diesel because of the mentioned reasons as rapid combustion capabilities, the thermal NO_x production is greater. Additionally, another factor for NO_x is a fuel–air ratio which plays a crucial role at greater than 0.5 equivalence ratio demonstrated in Fig. 4.3. Therefore, the lean combustion capability of hydrogen allows the operating of engines at lower power outputs which decreases the power density of the power plant (White et al. 2006).

Beyond the limit of unit equivalence ratio, seen in the "x" axis, the NO_x generation decreases rapidly. The figure also shows rich combustion capability of H_2 fuel. While

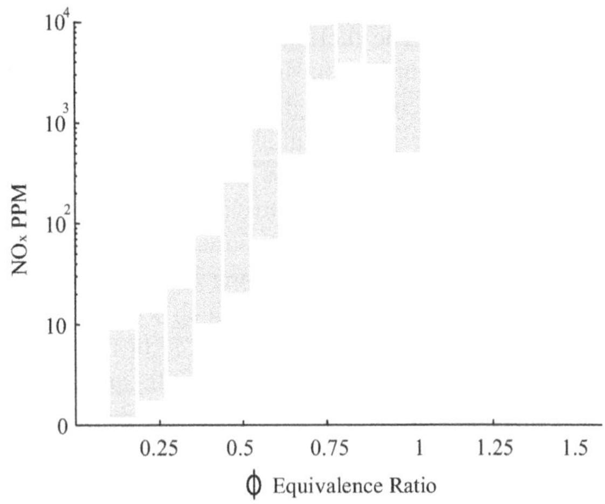

Fig. 4.3 NO_x production regarding to equivalence ratio for neat H_2 combustion, adapted from Inal et al. (2022)

low NO_x emission levels can be achieved with hydrogen lean combustion, the reduced NO_x emission levels can also be achieved with rich mixture combustion.

4.4 Hydrogen-Doped Operations in Dual-Fuel Engines

4.4.1 The Gasoline-Hydrogen Engines

In the gasoline engines, hydrogen addition can contribute to the power output energy share of the engine. Because of the less consumption of carbon-based fuels, the carbon dioxide emissions can be diminished by the help of hydrogen-assisted operations. Induction of hydrogen can be determined according to stoichiometric fraction or energy share equivalence in combustion.

The gasoline-hydrogen engines can be port fuel injection type or direct injection. As mentioned above in direct injection chapter, the direct injection strategy allows to reach higher power outputs than port fuel injection technology with an acceptable efficiency levels. The addition of hydrogen increases the combustion pressure and temperatures. Due to the increased combustion temperatures, NO_x emission level increases. However, there is a significant amount of reduction in carbon and hydrocarbon emissions in hydrogen-doped operations. Exhaust gas recirculation can be also implemented to hydrogen-doped operation of gasoline engines. In order to satisfy volumetric efficiency levels, hydrogen is directly injected into the combustion chamber while gasoline is injected at the port. Hydrogen helps the fast combustion of gasoline, increases the reactivity, and increases the efficiency. In particular, lean combustion capability of gasoline is significantly enhanced while satisfying good emission levels.

4.4.2 The LNG-Hydrogen Engines

Natural gas, having less carbon content than diesel, and relatively clean combustion capability look as a preferential and potential fuel for today and near future. It is also resistive to knocking which is very important in internal combustion engines. Hydrogen operation in internal combustion engines prones to knocking and the combination of hydrogen with a fuel which has a resistance to knocking can be considered as an adequate combination. Individually, natural gas has combustion disadvantages as slow burning and cycle-to-cycle fluctuations. Adding hydrogen to natural gas decreases the cyclic variation of natural gas combustion and increases the engine efficiency, together with advancing and shortening the combustion heat release phase (Zhang et al. 2020). The burning velocity of pure methane and the 50% each hydrogen-methane mixture comparison is demonstrated in the Fig. 4.4.

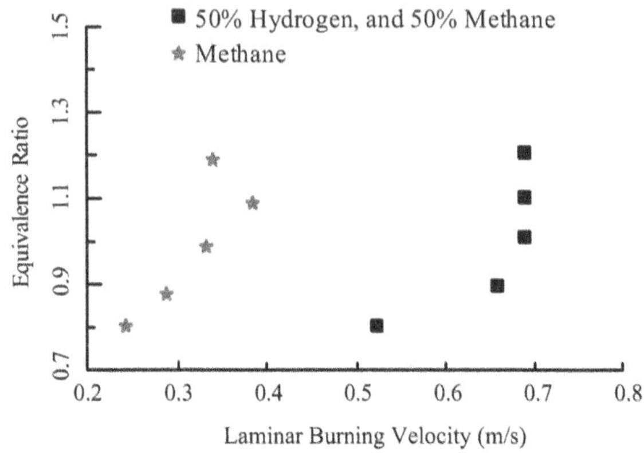

Fig. 4.4 Effect of hydrogen addition to natural gas on laminar burning velocity, adapted from Faizal et al. (2019)

The laminar burning velocity of natural gas is significantly increased with the help of the hydrogen addition. The mentioned problems above could be overcome by the hydrogen-doped operation. Additionally, natural gas powered engines have similarities with hydrogen-powered operations as like in spark-ignited engines.

4.4.3 Blending with Conventional Fuels (CNG)

Blending hydrogen with natural gas at certain fractions before the injection then usage of the blend in dual-fuel engine with the help of pilot diesel injection is one of the concept of hydrogen-doped operations. The sketch of the system is demonstrated in Fig. 4.5. This concept also called as tri-fuel operation that combination of H_2-CNG and diesel. CNG as an alternative fuel, with high calorific value, being resistive to knocking, relatively low CO_2 footprint comparing to diesel and adequate combustion capability make the fuel an alternative fuel (Chen et al. 2019). However, as a disadvantage of CNG, low flame propagation speed can be improved by mixing with a gas that has a high velocity of flame propagation (Huang et al. 2006). The fraction of hydrogen in the blend affects the emission formation as well as ignition delay and combustion properties. Adding hydrogen in to natural gas and afterward ignition of the blend with a pilot diesel fuel shows different combustion characteristics than standard dual-fuel operation. Ignition delay can increase because of the increased volume of hydrogen in the blend of natural gas. Particularly, at the advanced timings, more stabilized combustion occurs, and a part of diesel diffusive combustion can be shifted to premixed combustion phase in the concept of combustion (Lee et al. 2020).

Hydrogen addition in to the natural gas in a certain extend keeps the modifications away from the engine which is operated in dual-fuel operation natural gas/diesel. Diesel injection natural gas-diesel a-operation is a widespread practice in internal combustion engines. The operation can be carried out with the same way as injection

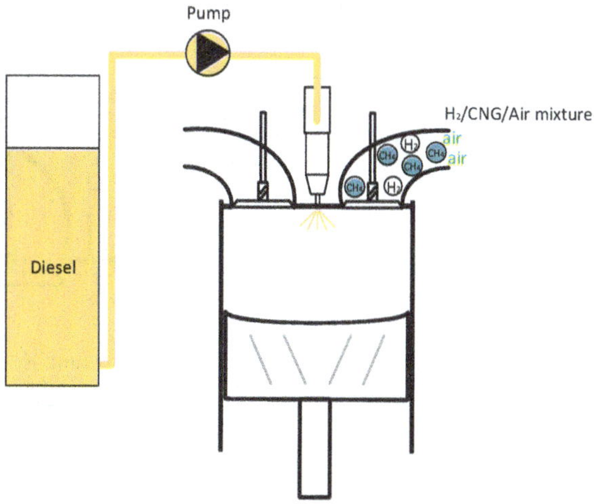

Fig. 4.5 Sketch of the system, blends CNG, and H_2 at intake manifold for IC

of diesel spray to ignite the mixture. Most of the combustion energies sourced from natural gas-hydrogen mixture.

4.4.4 The Diesel-Hydrogen Engines

Application of hydrogen fueled compressed ignition engines with diesel is a reasonable solution considering diesel fuel's low auto-ignition temperature. Diesel fuel is used to ignite the hydrogen air mixture. The diesel fuel is injected into the cylinder on to the mixture, while piston moving toward to top dead center. Although the combustion is initiated with diesel fuel, the main fuel is hydrogen, to be mixed with the air. The in-cylinder content, hydrogen/air mixture, can be prepared with carburation technique and induced in to the cylinders with intake manifold or ports. On the other hand, port injection technique proposes better combustion properties than carburation technique. Additionally, hydrogen can be introduced in to cylinder with injectors, as seen in Fig. 4.6, with a pressurized line from hydrogen tank with an assistance of pressure regulator. In order to avoid blow by of hydrogen, the hydrogen is injected after closing of exhaust valve during compression stroke. While hydrogen combustion is initiated by diesel pilot injection, compared to diesel combustion, it has quite advantageous combustion properties. According to studies, at the reduced loads, enhanced heat release rate combined with cylinder pressure increment, therefore, thermal efficiency improvement was observed with the induction of hydrogen in diesel engines (An et al. 2013).

The amount of induced hydrogen and the energy share ratio fuels do not remain same throughout engine load. The percentage of energy share and amount differ with the required loads. In the hydrogen diesel operations, the energy share of diesel can also be greater (Saravanan and Nagarajan 2010a) as well as the energy share

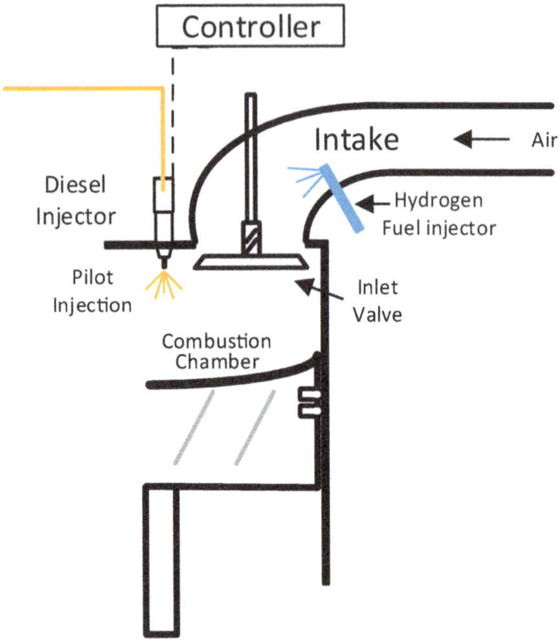

Fig. 4.6 Hydrogen injection sketch for diesel–H_2 operation

of hydrogen in some operations. The compression ratio has a crucial role on NO_x formation. Last but not least, hydrogen addition in diesel operations has an obvious favorable effect on smoke because of enhanced combustion performance (Shinde and Karunamurthy 2022).

4.5 The Discussion of Drawbacks and Advantages of H_2 Fueled Operations in Internal Combustion Engines

The hydrogen with its unique combustion and chemical properties needs a different processes than other conventional fuels. The pure H_2 fuel applications suffer from volumetric efficiency in carburation and port injection methods while direct injection surmounted this problem. It is obvious in the previous studies for neat and doped operations that the integration of H_2 fuel into the diesel operation leads the advantages as better combustion process, which comes with higher thermal efficiency, reduced carbon and soot emissions with higher NO_x emission. By the increase of H_2 energy share in the internal combustion engine, the NO_x emission level is prone to increase by the nature of NO_x generation mechanism, at increased temperatures, because of rapid combustion phenomenon near top dead center. There is instantaneous increase in both in-cylinder temperatures and pressure from ignition point to maximum pressure point in-cylinder. The in-cylinder temperature starts to decrease after max pressure point with the movement of the piston to BDC, the dramatic decrease causes to cease in NO_x reaction (Heywood 1988) consequently, NO_x emission remains as combustion product on account of small period of elevated in-cylinder temperatures. In order to minimize and optimize the NO_x generation, studies conducted by implementing exhaust gas recirculation (EGR) with different volumetric share in intake air (Wu and Wu 2012).

In order to control NO_x production, with EGR, exhaust gas has a high heat capacity because of vapor content and is used to decrease maximum combustion temperatures. Since the marine engines which have high compression ratios and high maximum pressures, comparing to land based or small-scale diesel engines, the in-cylinder temperatures are relatively high. Water injection strategy is another way to meet the NO_x emission levels. With a small amount of additional fuel consumption, around 2%, 87% of reduction in NO_x emission can be achieved with water injection strategy (Shinde and Karunamurthy 2022). Fuel injection strategy is also a type of NO_x reduction methods, however, it is not considered as successful as water injection method in the NO_x reduction techniques. Thanks to the EGR, which allows more fuel amount injection, more power densities can be achieved with the same size of engine. However, the power equivalent of the same engine is difficult to reach with pure hydrogen operations because of decreased volumetric efficiency levels. An additional increase in power output capability of engine is to provide more air into the engine. Because of the unintended combustion phenomena and operational deficiencies for instance; backfire, emissions, etc., supercharging with exhaust gas recirculation provide better power output capability and reduced emission levels regarding to NO_x can be achieved even in a port fuel injection engines (Verhelst et al. 2009). Because of the reason that the hydrogen can burn in a short period, the combustion can be considered as constant volume combustion due to its high burning velocity. The characteristic combustion feature of hydrogen allows better performance in terms of efficiency at reduced loads comparing to conventional fuels. So as to sum up, Table 4.2 shows a brief information about the hydrogen pros and cons.

On the other hand, the advance of spark-ignition near top dead center, the increased in-cylinder temperatures lead to loss more heat to the other mediums. It is obvious that increasing cylinder pressure leads to increase power output and thermal efficiency. Comparing the liquid fuel volumetric share, 1–2% of air in-cylinder, for an engine, the hydrogen volumetric share around 30% in port fuel injection operations. Although being low volumetric efficiency levels, brake thermal efficiency is slightly higher in hydrogen fueled operations; moreover, the higher brake thermal efficiency comes with reduced torque levels.

Table 4.2 Advantageous and disadvantageous properties of hydrogen

Advantages	Disadvantages
Zero-carbon content (reduced carbon emissions)	Volumetric efficiency (induction problems)
Rapid combustion characteristics	Storage problems
Better combustion process (@H_2 doped operations)	NO_x emissions
Higher thermal efficiency (@H_2 doped operations)	Piping and transportation of fuel
Higher heat loss (increased combustion temperature)	Low ignition energy (unintended combustion)
Wide range of stoichiometry combustion	Reduced output power
	Safety risk

4.6 Discussion of H_2 Fuel Applicability in Marine Engines

In the application for the marine engines, there are some investigations that the required output power of main engine of the ship is generated with neat hydrogen or combination of fuels.

A ship, roll-on/roll-off (RORO) type, is the subject of the research. The engine of the ship was operated with hydrogen, the efficiency could be achieved around 32%, and the relatively reduced efficiency is because of the high-temperature combustion-based cooling loses in the hydrogen engine (Seddiek et al. 2015). Additionally, that was observed as a result of the research that the operational load significantly affects cooling loses, while lean combustion operation has lower cooling loss than near stoichiometric combustions (Shudo et al. 2001). In the previous part, it was mentioned that the lean operation provides reduced NO_x levels because of the reduced in-cylinder temperatures. Thanks to the reduced combustion temperatures, low heat loss could be achieved in the hydrogen fueled operation. A large volume or engine size is required to meet the required power for a ship, with sufficient air induction and fuel. Another study was carried out with two fuels as natural gas and hydrogen combination, in a marine engine. A turbocharged operation had been conducted for provide a lean-burn combustion. The model/simulation method had been utilized to demonstrate the combustion phenomena in marine engines (Sapra et al. 2020). The results show parallel indicators with other hydrogen engine operations. Another natural gas engine study conducted by researchers (Leng et al. 2020) for a large-scale engine. The hydrogen enrichment to the operation had been implemented by pre-chamber initiated jet ignition for a large bore engine. The study also had been carried out with computational methods. The hydrogen-doped operation with a variable fraction and methane combustion had been modeled with the help of fluid dynamics software. While as the hydrogen fraction increases, the carbon emission decreases, and the NO_x emissions increase. The reasons are presented in the previous sections as thermal generation of NO_x. The reformed exhaust gas recirculation system is adopted in another marine natural gas engine together with hydrogen enrichment (Li et al. 2019). Together with the proposed system with the specified air–fuel ratios, the elevated rates of reformate exhaust gases the flame development, combustion duration, and the efficiency of the combustion could be improved.

In a two-stroke marine diesel engine, another hydrogen-assisted duel fuel operation had been conducted (Pan et al. 2014). Hydrogen enrichment had been implemented to the diesel engine with intake manifold integration, and a significant advantage had been seen at the idle speed when low-power requirement is needed that allows higher hydrogen ratio used in the combustion comparing the diesel engine consumption. The constraint can be derived from the injection technique, implemented on the engine. As an outcome of the study, significant reductions can be achieved when hydrogen contributes significantly in the inlet fuel energy portion. In other cases, as higher power outputs, the effect of hydrogen decreases as the fraction by energy share. Another study looked into the combination of hydrogen and heavy fuel oil combustion (HFO) in a marine diesel engine (Serrano et al. 2021). In the study, water injection was recommended as a NO_x emission reduction solution to meet the NO_x emission criterion. This research was also conducted in software environment for marine diesel engine. The research includes a modeling methodology for combustion prediction. With the dataset of HFO combustion for two-stroke low-speed (125 rpm) marine diesel engine with 16 MW power rate, the modeling investigation and validation were performed. The studies' findings are parallel with other internal combustion engine findings. Hydrogen shows similar combustion behavior in large-scale marine engines, however, operational parameters differ as maximum pressure and power. The temperature of the component naturally is affected by the combustion temperature. The injection techniques, mentioned in the previous section, were applied to marine engines. The expected drawbacks had been seen. Since the applicability of a new technology to onboard "marine engine in operation," some computational studies were carried out. They also show similar findings with small-scale engines.

The last but not least, storage of hydrogen has a critical role in the hydrogen applicability to the sea transportation in future. Storage of hydrogen is uneconomical both in storage with high pressure, low temperature, or with aid of metal hydrides and requires high-energy demand while storing or releasing (Andersson and Grönkvist 2019). When it is compared to other conventional fuels, stored with a liquid form, it can be said that the prominent problem in the use of hydrogen is storage (Inal et al. 2021). There are also alternative ways to store hydrogen as with ammonia, methanol, or LHG forms, the hydrogen content can be used by extracting the hydrogen of the compounds. Onboard applicability and the readiness level of storage technology and safety issues must be considered storage of hydrogen in future.

4.7 Conclusion

In all engines' dual-fuel operations, hydrogen addition increases the combustion speed, and high rate of heat release in a short period leads rapid pressure and temperature rise. They come with increased efficiency. Additionally, some studies showed that hydrogen addition increases the total heat release of the same amount of conventional fuel.

In the case of the hydrogen which is induced together with an air/fuel mixture, increasing the engine output power leads problem as backfire because of increased energy input to the engine. This is the challenge of overcoming the issue of high output and low emission trade-off in these systems. Marine engines, due to its power density, its operational parameters, and regarding in-cylinder temperature and pressures, are higher comparing to small-scale engines. In the increased power, outputs cause flame reflux from hotspots of the engine to the engine intake manifold during valve opening period. Therefore, control of valve timing is an essential control parameter to disable such kind of risk considering efficient induction period. Adjustment of valve timing can affect volumetric efficiency, particularly in retarded operations, but supercharging has a compensating effect on reduced volumetric efficiency, thus increased load can be achieved.

The marine engines required high-power output need boost of air in order to achieve high power. For the hydrogen operation, the engine fueled with hydrogen, to be introduced entirely into the engine market, it has to show the same advantages as like other conventional fueled engines such as high-power output, power density, low NO_x emissions, and high efficiency accompanied by its own advantages being near-zero carbon emissions. In order to provide the advantages together, there would be operations needed as implementing supercharging to achieve desired air/fuel ratios, low-temperature combustion so as to mitigate NO_x production. The boosting helps to reduce maximum temperatures in cylinder and enables to supply enough fuel into the engine while keeping air/fuel ratios at desired levels.

Future application in maritime sector and NO_x emissions is also as important as CO_2 emissions. For this purpose, any power generator should comply with the NO_x limitations governed by MARPOL. The treatment systems gain importance in this step. NO_x emissions can be reduced with a significant level with NO_x scrubbers. After all engine solutions to burn hydrogen, with a more effective and safely way, reduction of NO_x can be achieved thanks to the scrubbers. When the efficiency is on the issue, storage of hydrogen is another issue to be solved. Since the storage of hydrogen requires high energy demand or high heat energy to release from solid storage forms, the whole efficiency level of the system is highly dependent to the storage process, which will be one of the main barriers to hydrogen usage in maritime transportation.

References

Akal D, Öztuna S, Büyükakın MK (2020) A review of hydrogen usage in internal combustion engines (gasoline-Lpg-diesel) from combustion performance aspect. Int J Hydrogen Energy 45(60):35257–35268. https://doi.org/10.1016/j.ijhydene.2020.02.001

An H, Yang WM, Maghbouli A, Li J, Chou SK, Chua KJ (2013) A numerical study on a hydrogen assisted diesel engine. Int J Hydrogen Energy 38(6):2919–2928. https://doi.org/10.1016/j.ijhydene.2012.12.062

Andersson J, Grönkvist S (2019) Large-scale storage of hydrogen. Int J Hydrogen Energy 44(23):11901–11919. https://doi.org/10.1016/j.ijhydene.2019.03.063

Boretti A (2020) Hydrogen internal combustion engines to 2030. Int J Hydrogen Energy 45(43):23692–23703. https://doi.org/10.1016/j.ijhydene.2020.06.022

Chen H, He J, Zhong X (2019) Engine combustion and emission fueled with natural gas: a review. J Energy Inst 92(4):1123–1136. https://doi.org/10.1016/j.joei.2018.06.005

Deb M, Paul A, Debroy D, Sastry GRK, Panua RS, Bose PK (2015) An experimental investigation of performance-emission trade off characteristics of a CI engine using hydrogen as dual fuel. Energy 85:569–585. https://doi.org/10.1016/j.energy.2015.03.108

Faizal M, Chuah LS, Lee C, Hameed A, Lee J, Shankar M (2019) Review of hydrogen fuel for internal combustion engines. J Mech Eng Res Dev. 42(3):36–46. https://doi.org/10.26480/jmerd.03.2019.35.46

Herbinet O, Bartocci P, Grinberg DA (2022) On the use of ammonia as a fuel – a perspective. Fuel Commun. 11:100064. https://doi.org/10.1016/J.JFUECO.2022.100064

Heywood JB (1988) Internal combustion engine fundamentals. McGraw Hill, Inc., New York, USA

Huang Z, Zhang Y, Zeng K, Liu B, Wang Q, Jiang D (2006) Measurements of laminar burning velocities for natural gas-hydrogen-air mixtures. Combust Flame 146(1–2):302–311. https://doi.org/10.1016/j.combustflame.2006.03.003

Ianniello R, Belgiorno G, Di Luca G, Beatrice C, Di Blasio G (2021) Ethanol in dual-fuel and blend fueling modes for advanced combustion in compression ignition engines. Energy, Environ Sustain:5–27. https://doi.org/10.1007/978-981-16-0931-2_2

Inal OB, Deniz C (2020) Assessment of fuel cell types for ships: Based on multi-criteria decision analysis. J Clean Prod 265:121734. https://doi.org/10.1016/j.jclepro.2020.121734

Inal OB, Dere C, Deniz C (2021) Onboard hydrogen storage for ships : an overview onboard hydrogen storage. In: 5th international hydrogen technologies congress

Inal OB, Zincir B, Dere C (2022) Hydrogen as maritime transportation fuel: a pathway for decarbonization. In: Agarwal AK, Valera H (eds) Greener and scalable e-fuels for decarbonization of transport. Springer, US, pp 67–110. https://doi.org/10.1007/978-981-16-8344-2

Kim YY, Lee JT, Choi GH (2005) An investigation on the causes of cycle variation in direct injection hydrogen fueled engines. Int J Hydrogen Energy 30(1):69–76. https://doi.org/10.1016/j.ijhydene.2004.03.041

Kobayashi H, Hayakawa A, Somarathne KDKA, Okafor EC (2019) Science and technology of ammonia combustion. Proc Combust Inst 37(1):109–133. https://doi.org/10.1016/J.PROCI.2018.09.029

Lee C fon, Pang Y, Wu H, Hernández JJ, Zhang S, Liu F (2020) The optical investigation of hydrogen enrichment effects on combustion and soot emission characteristics of CNG/diesel dual-fuel engine. Fuel. 280(July):118639. https://doi.org/10.1016/j.fuel.2020.118639

Leng X, Huang H, Ge Q et al (2020) Effects of hydrogen enrichment on the combustion and emission characteristics of a turbulent jet ignited medium speed natural gas engine: a numerical study. Fuel 2021(290):119966. https://doi.org/10.1016/j.fuel.2020.119966

Li G, Long Y, Zhang Z et al (2019) Performance and emissions characteristics of a lean-burn marine natural gas engine with the addition of hydrogen-rich reformate. Int J Hydrogen Energy 44(59):31544–31556. https://doi.org/10.1016/j.ijhydene.2019.10.007

Mohammadi A, Shioji M, Nakai Y, Ishikura W, Tabo E (2007) Performance and combustion characteristics of a direct injection SI hydrogen engine. Int J Hydrogen Energy 32(2):296–304. https://doi.org/10.1016/j.ijhydene.2006.06.005

Pan H, Pournazeri S, Princevac M et al (2014) Effect of hydrogen addition on criteria and greenhouse gas emissions for a marine diesel engine. Int J Hydrogen Energy 39(21):11336–11345. https://doi.org/10.1016/j.ijhydene.2014.05.010

Sapra H, Godjevac M, De Vos P, Van Sluijs W, Linden Y, Visser K (2020) Hydrogen-natural gas combustion in a marine lean-burn SI engine: A comparative analysis of Seiliger and double Wiebe function-based zero-dimensional modelling. Energy Convers Manag 207(February):112494. https://doi.org/10.1016/j.enconman.2020.112494

Saravanan N, Nagarajan G (2008) An experimental investigation of hydrogen-enriched air induction in a diesel engine system. Int J Hydrogen Energy 33(6):1769–1775. https://doi.org/10.1016/j.ijhydene.2007.12.065

Saravanan N, Nagarajan G (2010a) An experimental investigation on hydrogen fuel injection in intake port and manifold with different EGR rates. Int J Energy Environ 1(2):221–248

Saravanan N, Nagarajan G (2010b) Performance and emission studies on port injection of hydrogen with varied flow rates with Diesel as an ignition source. Appl Energy 87(7):2218–2229. https://doi.org/10.1016/j.apenergy.2010.01.014

Seddiek IS, Elgohary MM, Ammar NR (2015) The hydrogen-fuelled internal combustion engines for marine applications with a case study. Brodogradnja. 66(1):23–38

Serrano J, Jiménez-Espadafor FJ, López A (2021) Prediction of hydrogen-heavy fuel combustion process with water addition in an adapted low speed two stroke diesel engine: Performance improvement. Appl Therm Eng 195:117250. https://doi.org/10.1016/j.applthermaleng.2021.117250

Shamun S, Belgiorno G, Di Blasio G (2020) Engine parameters assessment for alcohols fuels application in compression ignition engines. Energy, Environ Sustain:125–139. https://doi.org/10.1007/978-981-15-0418-1_8/COVER

Shinde BJ, Karunamurthy K (2022) Recent progress in hydrogen fuelled internal combustion engine (H2ICE) - a comprehensive outlook. Mater Today Proc 51(xxxx):1568–1579. https://doi.org/10.1016/j.matpr.2021.10.378

Shudo T, Nabetani S, Nakajima Y (2001) Analysis of the degree of constant volume and cooling loss in a spark ignition engine fuelled with hydrogen. Int J Engine Res. https://doi.org/10.1243/1468087011545361

Tsujimura T, Suzuki Y (2017) The utilization of hydrogen in hydrogen/diesel dual fuel engine. Int J Hydrogen Energy 42(19):14019–14029. https://doi.org/10.1016/j.ijhydene.2017.01.152

Verhelst S, Wallner T (2009) Hydrogen-fueled internal combustion engines. Prog Energy Combust Sci 35(6):490–527. https://doi.org/10.1016/j.pecs.2009.08.001

Verhelst S, Maesschalck P, Rombaut N, Sierens R (2009) Increasing the power output of hydrogen internal combustion engines by means of supercharging and exhaust gas recirculation. Int J Hydrogen Energy 34(10):4406–4412. https://doi.org/10.1016/j.ijhydene.2009.03.037

Wallner T, Nande AM, Naber JD (2009) Study of basic injection configurations using a direct-injection hydrogen research engine. SAE Tech Pap. 2(1):1221–1230. https://doi.org/10.4271/2009-01-1418

White CM, Steeper RR, Lutz AE (2006) The hydrogen-fueled internal combustion engine: a technical review. Int J Hydrogen Energy 31(10):1292–1305. https://doi.org/10.1016/j.ijhydene.2005.12.001

Wu HW, Wu ZY (2012) Investigation on combustion characteristics and emissions of diesel/hydrogen mixtures by using energy-share method in a diesel engine. Appl Therm Eng 42:154–162. https://doi.org/10.1016/j.applthermaleng.2012.03.004

Yip HL, Srna A, Yuen ACY et al (2019) A review of hydrogen direct injection for internal combustion engines: towards carbon-free combustion. Appl Sci. https://doi.org/10.3390/app9224842

Yu X, Wu H, Du Y, Tang Y, Liu L, Niu R (2016) Research on cycle-by-cycle variations of an SI engine with hydrogen direct injection under lean burn conditions. Appl Therm Eng 109:569–581. https://doi.org/10.1016/j.applthermaleng.2016.08.077

Zhang R, Chen L, Pan J, Wei H, Zhou L, Liu C (2020) Effects of direct-injected hydrogen addition on methane combustion performance in an optical SI engine with high compression-ratio. Int J Hydrogen Energy 45(4):3284–3293. https://doi.org/10.1016/j.ijhydene.2019.11.220

Zhen X, Li X, Wang Y, Liu D, Tian Z (2020) Comparative study on combustion and emission characteristics of methanol/hydrogen, ethanol/hydrogen and methane/hydrogen blends in high compression ratio SI engine. Fuel. https://doi.org/10.1016/j.fuel.2020.117193

Zincir B, Deniz C, Tunér M (2019) Investigation of environmental, operational and economic performance of methanol partially premixed combustion at slow speed operation of a marine engine. J Clean Prod 235:1006–1019. https://doi.org/10.1016/j.jclepro.2019.07.044

Chapter 5
Investigation and Examination of LNG, Methanol, and Ammonia Usage on Marine Vessels

Çağlar Karatuğ, Bulut Ozan Ceylan, Emir Ejder, and Yasin Arslanoğlu

Abstract This study aims to evaluate the use of LNG, methanol, and ammonia on ships as an alternative marine fuel. In this sense, firstly, the SWOT analysis is conducted, so the strengths and weak sides of the alternative fuels are determined. In the second step of the study, various criteria such as safety, cost, exhaust emission, global warming potential, sustainability, storage, and technical competence are specified, and the alternative fuels are analyzed with the TOPSIS method based on the identified criteria. As a result of the obtained judgments from the marine experts, the safety of fuel, its global warming potential, and its storage feature is determined as the most influential comparison weights. In addition, ammonia is determined as the best fuel option based on the 2.92 similarity value while values of LNG and methanol are calculated 2.21 and 2.18, respectively. Then, a sensitivity analysis where the various cases were created by improving the weights of criteria by 25% and applying the same weight value for each criterion is conducted to reveal the criticality of the criterion weighting. According to this analysis, it is observed that the analysis is highly sensitive to the global warming potential criteria. In line with this information, beneficial and significant key findings to policy-makers, stakeholders, and maritime companies are presented from the perspectives of short-term and long-term emission reduction strategies.

Keywords Maritime · Alternative marine fuel · SWOT · TOPSIS · Sensitivity analysis

Ç. Karatuğ (✉) · B. O. Ceylan · E. Ejder · Y. Arslanoğlu
Maritime Faculty, Istanbul Technical University, 34940 Istanbul, Turkey
e-mail: karatug@itu.edu.tr

B. O. Ceylan
e-mail: bceylan@bandirma.edu.tr

E. Ejder
e-mail: ejder18@itu.edu.tr

Y. Arslanoğlu
e-mail: arslanoglu@itu.edu.tr

B. O. Ceylan
Department of Marine Engineering, Bandirma Onyedi Eylül University, 10200 Balikesir, Turkey

B. Zincir et al. (eds.), *Decarbonization of Maritime Transport*, Energy, Environment, and Sustainability, https://doi.org/10.1007/978-981-99-1677-1_5

Abbreviations

AHP	Analytic hierarchy process
LCA	Life cycle assessment
BOG	Boil-off gas
LNG	Liquefied natural gas
CI	Compression ignition
LD	Light duty
DICI	Direct injection compression ignition
DI	Direct injection
ECA	Emission control area
MCDM	Multi-criteria decision-making
GHG	Greenhouse gas
MDO	Marine diesel oil
HFO	Heavy fuel oil
SWOT	Strengths, weaknesses, opportunities, and threats
IMO	International maritime organization
TOPSIS	Technique for order of preference by similarity to ideal solution

5.1 Introduction

Maritime transportation is a significant part of the global cargo supply chain and provides 80% of world trade (UNCTAD 2017; Elidolu et al. 2022). Eight billion tons of international trade goods have been carried by shipping every year (Du et al. 2011). Ensuring high-volume transportation, high-powered ship diesel engines with various integrated complex systems were used in the ship engine rooms (Ceylan et al. 2022a, b). As a result of the high fuel consumption required by these high-power diesel engines during ship transportation, exhaust gas emissions are generated. The International Maritime Organization (IMO) which steers the shipping sector has been stated in the 3rd greenhouse gas study that approximately 300 million tons of fuel which are mostly heavy fuel oil (HFO) were consumed annually (IMO 2014). As a result of the combustion of the HFO, serious amounts of pollutants such as CO_2, SO_x, and NO_x have been emitted into the atmosphere (Ceylan et al. 2022b; Karatuğ and Arslanoğlu 2022). It has been presented by the IMO that the portion of the maritime sector in global anthropogenic emissions is 2.89% in 2018 (IMO 2020).

Due to the harmful effects of these types of emission gases, IMO introduced some emission-related rules (IMO 1997). Additionally, the IMO defined the decarbonization strategy of the maritime sector (IMO 2018) to reduce pollution caused by ships. Accordingly, it is aimed to reduce total annual greenhouse gas (GHG) emissions by at least 50% by 2050, compared to 2008. To decrease the amounts of SO_x, after 1 January 2020, the sulfur content limit of the fuel is reduced from 3.50% m/m to 0.50% m/m for ships that navigate on open seas as shown in Fig. 5.1, while it is

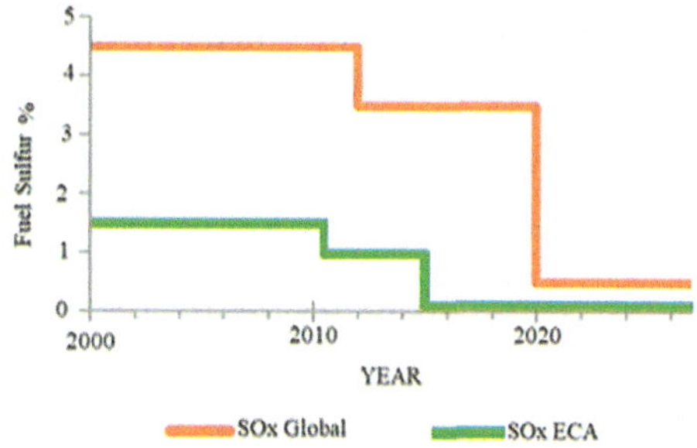

Fig. 5.1 IMO sulfur limits (IMO 2020)

determined as 0.01% m/m for ships where navigate in emission control areas (ECA). Also, some limits for the NO_x were defined accordingly. These strict rules force shipping companies and operators to research emission reduction approaches and implement these methods in their ships. In this sense, different research areas such as the use of alternative energy sources (Karatuğ and Durmuşoğlu 2020), exhaust gas treatment applications (Deng et al. 2021), and investigation of green alternative fuels (Deniz and Zincir 2016) stand out in the maritime sector.

In the 4th IMO GHG study, it was presented that HFO is still the most widely used marine fuel with 79%. On the other hand, it is understood that the use of marine diesel oil (MDO) and liquefied natural gas (LNG) as main fuels in the world fleet has increased with the last sulfur limitation that came into effect in 2020. It was also stated in the study that methanol is the 4th most common marine fuel. In addition to these fuels, ammonia is one of the promising fuels for the maritime industry due to its carbon-free structure and its compliance with the decarbonization target determined by IMO (Kim et al. 2020a).

Each of the specified alternative marine fuels has both different advantages and disadvantages. In this study, LNG, methanol, and ammonia, which are recently been intensively researched as alternative marine fuels, were examined with strengths, weaknesses, opportunities, and threats (SWOT) analysis. Then, some criteria to be important for the preference of the alternative marine fuel have been determined and analyzed by the technique for order of preference by similarity to ideal solution (TOPSIS) method which is one of the most common multi-criteria decision-making (MCDM) methods. Lastly, a sensitivity analysis was performed to observe the importance level of each criterion for the similarity values of each alternative marine fuel. As a result of the analysis, ammonia is found as the best fuel alternative, while the most critical comparison criterion is stated as global warming potential. The LNG has currently practical implementation, so its technical competency is superior to methanol. However, the closeness of the similarity values of the LNG and methanol could be interpreted as methanol can be an alternative to LNG when its technical competence is sufficiently developed.

For researchers interested in this field and maritime companies, the proposed methodology enables both firstly, to evaluate the advantages and disadvantageous sides of the alternatives within the SWOT analysis and secondly, to determine the best option according to the general intention of the expert consortium. In addition, different from the relevant literature, the inclusion of SWOT analysis in the proposed approach has enhanced the influence level of the selection of fuel alternatives via methodology by handling each fuel option from different points of view.

5.1.1 Literature Review

There are some studies about alternative fuels in the literature. They have been either examined individually or analyzed comparatively. Pucilowski et al. (2017) investigated the methanol-fueled heavy-duty direct injection compression ignition (DICI) engine combustion characteristics by using the start of injection effect. Zincir et al. (2019a) use an experimental approach to investigate how intake temperature affects the low load limits of partially premixed combustion of the same alternative fuel (methanol). Iannaccone et al. (Iannaccone et al. 2020) evaluated LNG under some environmental and safety factors and proposed that compared to the diesel-fueled system, the LNG system was 41% and 61% more effective in terms of environment and safety, respectively. Ammar (2019) evaluated the application of a methanol dual-fuel engine for a container ship from an environmental and economic perspective. He presented that the dual-fuel system would provide savings in 12 years, while reductions occurred in emission releasing. Hansson et al. (2020) evaluated ammonia as a future marine fuel. They stated that although it is a potential fuel due to its low environmental damage, significant technical applications should be structured and developed.

Perčić et al. (2021) carried out the economic analysis of different alternative marine fuels using the life cycle assessment (LCA) method. They stated that although methanol is the most cost-effective fuel, the necessary system bunkering infrastructure should be developed. Al-Breiki and Bicer (2020) realized the energy and exergy analysis of the three fuels studied in the study and calculated boil-off gas (BOG) ratios of them. They found that the most loss of fuel occurs in LNG systems. McKinlay et al. (2021) calculated that the maximum power demand per voyage is 9270 MWh, based on raw shipping data. Accordingly, ammonia, hydrogen, and methanol systems that can provide this power have been designed, and these systems are examined under sub-headings: storage infrastructure, desired design range, and both. Xing et al. (2021) discussed future alternative marine fuel options and presented that renewable methanol is the most promising fuel option globally, and ammonia is useful in domestic and short-sea shipping.

Wan et al. (2015) carried out a hybrid methodology based on SWOT analysis and the analytic hierarchy process (AHP) to evaluate the development of LNG-fueled ships in the inland waters of China. Some studies, on the other hand, examined duel fuel or more fuel blends instead of focusing on a single fuel. Di Blasio et al. (2017) used a dual fuel (methane-diesel) for the investigation of the performance, emissions, and particle size distributions of light duty (LD) diesel engine. Fraioli et al. (2017) carried out another dual-fuel study. They investigate the combustion of methane and diesel fuel mixture on LD diesel engines by utilizing multidimensional simulations. Balasubramanian et al. (2021) used waste cooking oil biofuel and diesel blends to investigate the emission, performance, and combustion of a single-cylinder compression ignition (CI) engine. Kumar et al. (2021) carried out diesel and methanol fuel mixture combustion, performance, and emission analysis in CI Engine. Shamun et al. (2018) carried out performance and emissions analysis of diesel, biodiesel, and ethanol blends in a single-cylinder LD CI engine. With a similar approach, Belgiorno et al. (2018) investigate the performance of diesel, gasoline, and ethanol blends in an LD CI engine.

The rest of the paper is organized as follows. The brief information for specified marine fuels, SWOT analysis, TOPSIS method, and methodological approach is presented in Sect. 5.2. The case study is conducted in Sect. 5.3. In the final, the key findings of the paper are presented in Sect. 5.4.

5.2 Materials and Methodology

5.2.1 Alternative Marine Fuels

The utilization of alternative marine fuel sources instead of HFO is a significant method to reduce emissions. There is a strong trend toward the use of alternative fuels with the intent of reducing the environmental impacts of shipping. Today, many researchers are conducting various scientific research on this current issue (Hansson et al. 2019; Paulauskiene et al. 2019; Perčić et al. 2020; Lunde Hermansson et al. 2021; Chu et al. 2019). Within the scope of this study, brief information about the use of LNG, ammonia, and methanol as marine fuels has been given in this section.

5.2.1.1 Liquefied Natural Gas

LNG is an environmentally friendly fuel type in the gas state that has been started to use as the main energy source of many vessels. Additionally, it can be used with other fuels in dual-fuel engines (Bilgili 2021). With the recent international restrictions, developing technology, and maritime field economics, LNG is becoming attractive marine fuel. LNG provides a 25% CO_2 reduction compared to HFO (Iannaccone et al. 2020). After the combustion process, a low rate of NO_x and PM has been produced by LNG usage compared to the HFO and also, and it is not released SO_x (Kim et al. 2020b). Moreover, LNG has a fair price when compared to other alternative marine fuels. However, LNG also has some risks, for instance, it must be stored in very well insulated tanks and needs more storage space. Therefore, this may cause additional costs. The other disadvantage of LNG is that this fuel alone cannot comply with the international requirements of 50% CO_2 reduction (DNV GL 2019).

5.2.1.2 Methanol

The other alternative marine fuel is methanol. With the IMO 2020 regulations, it can be used to reduce emissions. Methanol, CH_3OH, is a simple oxygenated hydrocarbon that ranks in the top five of the most traded chemicals in the world (Verhelst et al. 2019; Zincir et al. 2019b). It is a liquid and a sulfur-free corrosive fuel. It easily burns with CO_2 and H_2O, emitting no SO_x and low NO_x and PM. Methanol can be obtained from natural gas or coal. The simplest alcohol, methanol, has a low flash point, and it is a very risky marine fuel due to toxicity. It is a highly flammable gas because its calorific value has been calculated as 20,000 MJ/t (Bilgili 2021; Gilbert et al. 2018). Methanol is used in some successful marine trials and commercial projects as fuel (Liu et al. 2019). It has a low flash point at 11 °C, which does not comply with the safety of life at sea convention of IMO. However, according to the studies, a double-wall design of methanol components can solve this problem (Ammar 2019).

5.2.1.3 Ammonia

Ammonia (NH_3) is an increasingly studied, sustainable fuel for global use in future. It is a carbon-free alternative fuel that is utilized in many sectors such as healthcare, plastics, textiles, cosmetics, nutrition, and electronics (Hansson et al. 2020). Additionally, ammonia can be used in diesel engines, gas turbines, and fuel cells (Kim et al. 2020b).

Ammonia includes 1 nitrogen and 3 hydrogen atoms. In addition to its carbon-free structure, it is also a sulfur-free molecule. Therefore, combustion products of ammonia do not contain CO, CO_2, or SO_x emissions. After the ignition, only water and nitrogen products are formed. Ammonia is liquefied by 10 bar pressure at room temperature, or by −33 °C atmospheric pressure. Ammonia, which produces around 175 million tons per year worldwide, compared to liquid hydrogen, transportation, and pipeline transfer technology, is advanced for the current industry (MacFarlane et al. 2020). It is considered a strong alternative to hydrogen fuel (Bilgili 2021). However, ammonia is hardly ignited fuel, and compared to the other alternative fuels, it is toxic for both humans and the environment. Additionally, considering the fuel system and its components, ammonia is a corrosive substance (Zincir 2020).

5.2.2 SWOT Analysis

SWOT analysis can be performed with the analysis of the current situation as a whole and its internal and external environment (Olabi et al. 2022). This analysis aims to reveal the current situation, determine priorities, and identify strategic issues for progress and development. Analyzing the internal environment is a method that allows revealing the opportunities and threats by analyzing the external environment while identifying the strengths and weaknesses (Stavroulakis and Papadimitriou 2017). Strengths are the capabilities and assets that enable the situation to gain an advantage over its competitors and are both practical and efficient. On the other hand, weaknesses refer to situations where it is more inadequate, inefficient, ineffective, and powerless than its competitors. Variables consist of technological, social, cultural, economic, and global environmental elements, and the positive results of these elements for current situations are opportunities. Threats include situations that occur due to the change in external environmental factors, which may prevent the business from continuing its existence or cause it to lose its competitive advantage (Hossain et al. 2017; Al-Haidous et al. 2022; Efe et al. 2022). The SWOT analysis identifies critical internal and external factors, allowing weaknesses to be reduced and strategic planning for threats to be created effectively while taking strengths and opportunities into account.

5.2.3 TOPSIS Method

The TOPSIS method, based on the idea of approaching the ideal solution, allows the identification or selection of the optimal choice in any situation requiring decision-making by computing the positive and negative ideal solution distances (Wang et al. 2022). The method can handle very constrained decision criteria and effectively solve the decision problem. In addition, the TOPSIS method enables the creation of a standardized matrix, often derived from expert experience, in determining weights for criteria. TOPSIS facilitates analysis by assigning functions to evaluations and digitizing them, allowing for joint decision-making in problems involving many criteria and alternatives (Yang et al. 2022). The most prominent feature is that the importance weights of the criteria are different from each other. It is a convenient method for solving problems effectively and thus provides the ability to deal with uncertainty in decision-making (Bin Din et al. 2022; Zhang et al. 2022; Chrysafis et al. 2022). The algorithmic phases of the TOPSIS methods were presented as follows:

Step 1: The decision matrix is an $M \times N$ dimensional matrix created by the decision-maker after the decision options, and evaluation criteria are determined.

$$a(ij)_{M \times N} \tag{5.1}$$

where N and M are the numbers of decision options and evaluation criteria.

Step 2: A standard decision matrix (normalized matrix) is created. If the value of any element of the decision matrix is 0, the value of the relevant component in the standard decision matrix will also be 0. The normalized decision matrix can be defined as follows:

$$a_{ij} = \frac{a_{ij}}{\sqrt{\sum_{i=1}^{M} (xij)2}} \tag{5.2}$$

Step 3: A weighted standard decision matrix is created. Weight values for evaluation criteria are determined. A weighted standard decision matrix is formed by multiplying the elements of the matrix with their respective weight values.

$$X_{ij} = a_{ij} \times w_{ij} \tag{5.3}$$

$$w_{ij} = \frac{w_j}{\sum_{j=1}^{N} w_j} \tag{5.4}$$

$$\sum_{j=1}^{N} w_j = 1 \tag{5.5}$$

Step 4: Positive ideal and negative ideal solution values are obtained.

$$S^* = max_{i=1}^{M} X_{ij} \tag{5.6}$$

$$S^- = min_{i=1}^{M} X_{ij} \tag{5.7}$$

Step 5: The distance values to the positive ideal and negative ideal solution values are obtained.

$$d^* = \sqrt{\sum_{j=1}^{N} (X_{ij} - S^*)2} \tag{5.8}$$

$$d^- = \sqrt{\sum_{j=1}^{N} (X_{ij} - S^-)2} \tag{5.9}$$

Step 6: The distances of each alternative from the positive and negative perfect solutions are calculated.

$$S_{SV} = \frac{d^-}{d^* + d^-}, i = 1, 2, 3, \ldots, n \tag{5.10}$$

where $0 \leq S_{SV} \leq 1$ is the share of the distance to the ideal solution in the total distance. Accordingly, S_{SV} decision options close to 1 are preferred primarily.

5.2.4 *Methodical Approach*

While alternative fuels are a major topic in the marine industry, there are diverse perspectives on which fuel would be the most beneficial. In this study, frequently used LNG, methanol, and ammonia fuels in the literature were evaluated, and the best alternative was determined. For this purpose, the methodological approach of the study was designed. In this framework, the methodological approach of the study consists of two steps. The first step includes the SWOT analysis of specified alternative marine fuel types. The second stage of the study continues with the help of the data obtained by revealing the strengths-weaknesses and threats-opportunities of the fuels. This step of the study includes the evaluation of alternative marine fuels with the MCDM method. To conduct analysis, some criteria were determined based on SWOT analysis conducted and research on relevant literature (Hansson et al. 2020, 2019; Balcombe et al. 2019; Inal et al. 2022; Inal and Deniz 2020; Andersson et al. 2020). The TOPSIS approach was used to analyze fuel options based on the criteria such as safety, cost, exhaust emission, global warming potential, sustainability, storage, and technical competence. Experts were asked to score the importance of each criterion and three fuel types based on these criteria. Finally, the best fuel alternative was determined once the score was received. The methodical approach of the study was demonstrated in Fig. 5.2.

Engineers and academicians who have worked on ships using various fuel types were employed as experts in the study. Table 5.1 shows the profiles of the experts who participated in the study.

5.3 Case Study

In this paper, firstly, the specified alternative fuels were examined by SWOT analysis. Thus, the advantageous and disadvantageous aspects were determined, and the main criteria to be considered in the selection of alternative fuels were revealed. Secondly, a useful strategy to select the most suitable alternative marine fuel is presented. The LNG, methanol, and ammonia have been analyzed based on some criteria such as safety, cost, exhaust emission, global warming potential, sustainability, storage, and technical competence through the TOPSIS method.

5.3.1 *SWOT Analysis of Alternative Marine Fuels*

The SWOT analysis was performed using some studies from the literature, and the results were used to identify the strengths, weaknesses, opportunities, and threats of alternate marine fuels. The obtained results are presented in Appendix 5.1.

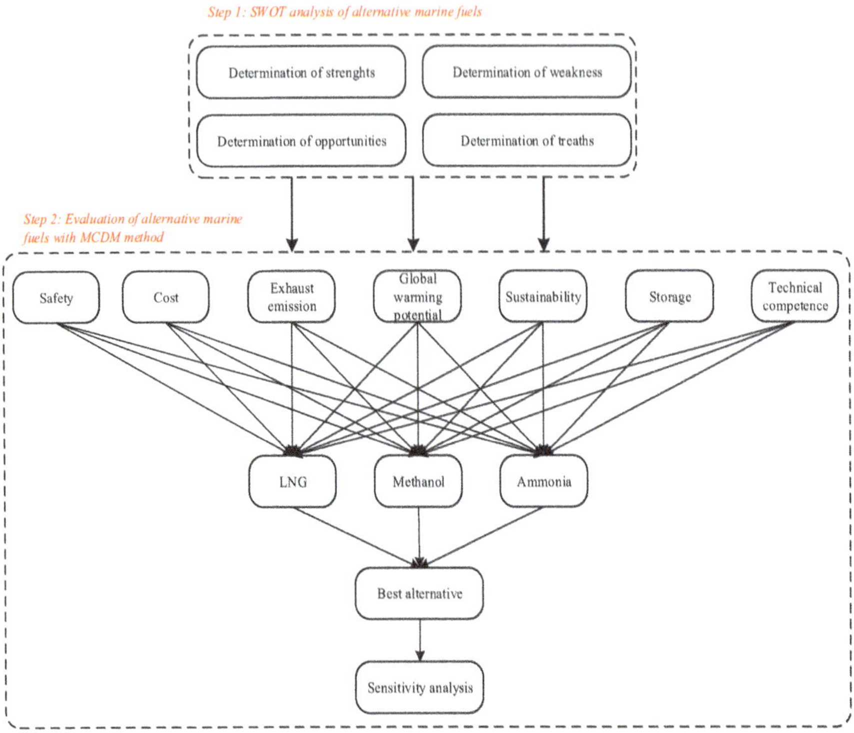

Fig. 5.2 Methodical approach of the study

Table 5.1 Expert profiles of the study

Experts	Ship experience	Current position
Expert 1	Chief Engineer	Shipping Company-Oceangoing Chief Engineer
Expert 2	Chief Engineer	Shipping Company-Oceangoing Chief Engineer
Expert 3	Chief Engineer	Shipping Company-Oceangoing Chief Engineer
Expert 4	First Engineer	National Maritime Authority-Port State Control Officer
Expert 5	First Engineer	University-Academician
Expert 6	First Engineer	Shipping Company-Oceangoing First Engineer
Expert 7	Second Engineer	University-Academician

5.3.2 TOPSIS Application

In the second part of the methodology, the specified alternative marine fuels were analyzed by TOPSIS. The alternative fuels were evaluated based on some significant criteria related to alternative selection such as safety, cost, exhaust emission, global warming potential, sustainability, storage, and technical competence.

Table 5.2 Performance scores for criteria

Criteria	Performance scores	
Importance of criteria	1-Worst	5-Best
C1- Safety	1-Worst	5-Best
C2- Cost	1-Most expensive	5- Most economic
C3- Exhaust emission	1-Worst	5-Best
C4- Global warming potential	1-Worst	5-Best
C5- Sustainability	1-Worst	5-Best
C6- Storage	1-Worst	5-Best
C7- Technical competence	1-Worst	5-Best

The analysis was realized based on the scores received by marine experts who are marine engineers or academicians in the maritime field. Four of the marine experts work on ships, and they have operational experience with different types of marine fuel. Two of the marine experts have sea service experience and currently, work at the university. One of the marine experts is the first engineer and works as the port state control officer. In the first stage, marine experts were asked to judge the criteria and criterion weights based on the information presented in Table 5.2.

The decision matrix was formed by taking the average of the scores obtained from the experts. The constituted decision matrix is as in Table 5.3.

The aggregated decision matrix was normalized using Eq. 5.2. Then, a weighted normalized decision matrix was created by introducing weights of each criterion to normalized values. It is presented in Table 5.4.

Based on values in Table 5.5, the best S^* and worst S^- alternatives are determined and presented in Table 5.6.

The next step of the analysis is the calculation of distances between the target alternative and both the best alternative and worst alternative. These calculations were realized using Eqs. 5.8 and 5.9. After calculation of the distances, the similarity value S_{SV} to the worst alternative for each alternative was determined. While $S_{SV} = 1$

Table 5.3 Decision matrix

Criteria	Weight	Alternatives		
		LNG	Methanol	Ammonia
C1	3.86	3.71	3.14	2.71
C2	3.00	3.29	3.29	2.14
C3	3.29	3.14	3.14	4.00
C4	3.86	2.00	2.57	4.29
C5	3.14	3.14	3.14	2.86
C6	3.43	2.43	3.86	3.14
C7	3.14	3.71	2.71	2.00

Table 5.4 Weighted normalized decision matrix

Criteria	Alternatives		
	LNG	Methanol	Ammonia
C1	2.57	2.18	1.88
C2	1.93	1.93	1.26
C3	1.73	1.73	2.20
C4	1.43	1.84	3.07
C5	1.87	1.87	1.70
C6	1.50	2.39	1.95
C7	2.33	1.70	1.25

Table 5.5 Best and worst alternatives

Criteria	S^*	S^-
C1	2.57	1.88
C2	1.93	1.26
C3	2.20	1.73
C4	3.07	1.43
C5	1.87	1.70
C6	2.39	1.50
C7	2.33	1.25

Table 5.6 Determination of best alternative

Alternatives	d^*	d^-	S_{SV}
LNG	1.92	1.45	2.21
Methanol	1.51	1.31	2.18
Ammonia	1.52	1.76	2.92

means that the alternative is the best solution, $S_{SV} = 0$ represents that the alternative is the worst solution. The best and worst distances of each alternative and their similarity values are presented in Table 5.6.

5.3.2.1 Sensitivity Analysis

Sensitivity analysis is an important application for MCDM studies. It provides an important projection of how effective the identified criteria are on the result obtained. In particular, the scores obtained in an MCDM application developed based on expert opinion are subjective, no matter how much they are obtained from experts that work

Table 5.7 Formed cases for sensitivity analysis

Case	Description	Weight of criterion						
		C1	C2	C3	C4	C5	C6	C7
Base case	Base condition weighting	3.86	3.00	3.28	3.85	3.14	3.42	3.14
Equal weights	Equal weighting	**3.39**	**3.39**	**3.39**	**3.39**	**3.39**	**3.39**	**3.39**
Case 1	C1 + 25% weighting	**4.83**	3.00	3.28	3.85	3.14	3.42	3.14
Case 2	C2 + 25% weighting	3.86	**3.75**	3.28	3.85	3.14	3.42	3.14
Case 3	C3 + 25% weighting	3.86	3.00	**4.11**	3.85	3.14	3.42	3.14
Case 4	C4 + 25% weighting	3.86	3.00	3.28	**4.83**	3.14	3.42	3.14
Case 5	C5 + 25% weighting	3.86	3.00	3.28	3.85	**3.93**	3.42	3.14
Case 6	C6 + 25% weighting	3.86	3.00	3.28	3.85	3.14	**4.29**	3.14
Case 7	C7 + 25% weighting	3.86	3.00	3.28	3.85	3.14	3.42	**3.93**

in the relevant field. Therefore, the results could vary in the evaluation conducted by a different consortium of experts (Inal et al. 2022). At this point, the sensitivity analysis reveals the effect of the changes in the weights of the criteria on the result obtained and enables the determination of critical criteria. In the sensitivity analysis, the various cases were created by improving the weights of criteria by 25% and applying the same weight value for each criterion. The formed cases for sensitivity analysis and weights of criterion for each case are illustrated in Table 5.7.

The same calculations with the base case have been made for each formed case. The effects of changes on the distance to best and worst alternatives and similarity value were observed. The changes that occurred as a result of the calculations made within the scope of the sensitivity analysis are illustrated and presented in Fig. 5.3.

The rank of the preference of the specified alternative marine fuels was mostly observed as Ammonia > LNG > Methanol. However, it should be underlined that LNG and methanol have generally close similarity values in cases created. It is observed that the similarity value of the methanol is raised with increasing the weight of the storage criteria since the storage of methanol could be achieved with a small arrangement for the existing ships. For ammonia, the global warming potential is revealed as the most dominant criterion. The increase of this criterion weighting by 25% in case 4 perceptibly raised the similarity value of ammonia. This situation is directly related to ammonia's carbon-free structure.

5.4 Conclusions

The importance of reducing emissions from the maritime sector is growing every day. Using alternative marine fuels on ships offers excellent benefits for shipping companies in terms of reducing pollution. Furthermore, choosing the appropriate

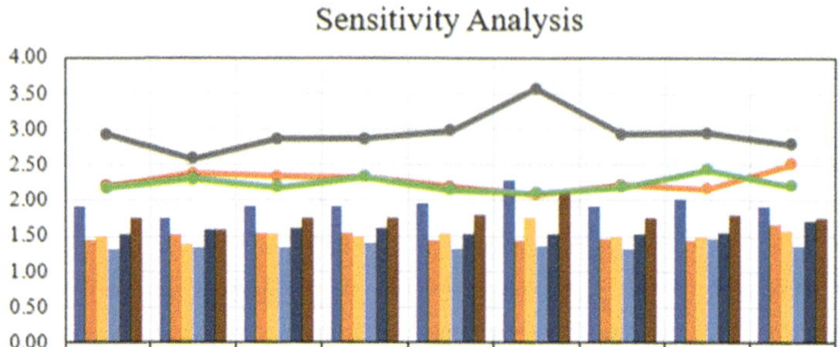

	Base case	Equal weights	Case 1	Case 2	Case 3	Case 4	Case 5	Case 6	Case 7
LNG d*	1.92	1.75	1.92	1.92	1.95	2.28	1.92	2.03	1.92
LNG d⁻	1.45	1.52	1.54	1.54	1.45	1.45	1.46	1.45	1.66
Methanol d*	1.51	1.41	1.54	1.51	1.55	1.77	1.51	1.51	1.58
Methanol d⁻	1.31	1.34	1.33	1.4	1.31	1.35	1.32	1.47	1.35
Ammonia d*	1.52	1.58	1.61	1.6	1.52	1.52	1.52	1.55	1.72
Ammonia d⁻	1.76	1.58	1.76	1.76	1.8	2.15	1.76	1.79	1.76
LNG Similarity	2.21	2.39	2.35	2.34	2.2	2.09	2.22	2.17	2.52
Methanol Similarity	2.18	2.3	2.19	2.33	2.15	2.11	2.19	2.44	2.21
Ammonia Similarity	2.92	2.58	2.86	2.86	2.98	3.56	2.92	2.94	2.79

Fig. 5.3 Results of sensitivity analysis

alternative fuel for both short-term and long-term investments may have significant benefits for the shipping industry.

In this study, a framework has been presented to determine the best alternative marine fuel option for marine vessels. LNG, methanol, and ammonia were considered throughout the analysis as alternative marine fuels. In the first part of the study, the stated fuels were analyzed by the SWOT analysis method. Thus, the advantages and disadvantages of these fuels have been identified. In addition, a process to determine the criteria that are important during the preference of alternative marine fuel use on board has been conducted with the SWOT analysis and research on the relevant literature. Within the scope of the methodology, safety, cost, exhaust emission, global warming potential, sustainability, storage, and technical competence were considered, and specified fuels were examined based on these criteria by the TOPSIS approach. Some marine professionals who work as marine engineers at various levels on board or academicians working in maritime education were asked to score criteria to conduct the analysis. The obtained judgments are analyzed, and the best option was determined. A sensitivity analysis was carried out to reveal the effect of the criterion weighting for the alternatives, and key findings were presented. The main outcomes of the study are as follows:

- Among the comparison criteria, the safety, global warming potential of the fuel, and its storage are found most important criteria.
- Among the fuel alternatives, ammonia is determined as the best alternative, while it is observed that LNG and methanol shared highly close similarity values as a result of the TOPSIS analysis.
- Although ammonia is a very promising fuel option for the maritime industry to eliminate ship-borne pollutants, there are some essential issues to be dealt with about its application.
- LNG has currently superiority within more technical competence, and the sector is familiar with its usage since complying with sulfur restrictions while methanol could be more adapted than its current status with a small arrangement in existing ships in the recent future.
- As a result of the sensitivity analysis, it is understood that the conducted analysis is very sensitive to the changing of C4.
- Within the scope of the study, a hybrid methodology that includes SWOT analysis and a multi-criteria decision-making approach is presented to determine the best alternative fuel option. Compared to the relevant literature, the inclusion of SWOT analysis in the methodology has strengthened the accuracy and effectiveness of the approach.
- For researchers interested in this field and maritime companies, the proposed methodology enables both firstly, to evaluate the advantages and disadvantageous sides of the alternatives within the SWOT analysis and secondly, to determine the best option according to the general intention of the expert consortium.

This study allows a beneficial framework for maritime companies to determine suitable alternative marine fuels for their ships in the fleet. On the other hand, the proposed methodology has a limitation in which it may be shaped according to the desire and intention of the expert consortium because it covers subjective judgments about the specified fuel options, comparison criteria, and their importance weights. In future studies, this study will extend by including more alternative marine fuel options and realizing analysis with various MCDM strategies. Also, we are planning to evaluate ammonia more deeply in future studies by considering ammonia fuel options such as those produced from natural gas or electrolysis based on renewable electricity and for use in fuel cells.

Appendix 5.1: SWOT Analysis of Marine Alternative Fuels

Types of fuels	Strengths	Weaknesses	Opportunities	Threats
LNG	• It reduces SO_x and PM emissions by 90–95% • It can reduce CO_2 emissions by approx. 25% • There are regulations for the use of LNG fuel • The reserve estimate is more than 250 years • It is cheaper than fossil fuels • It is non-explosive in a liquid state • It is not toxic • It is not corrosive • Safe gas operation	• It has a lower energy density than fuel oils • Larger volumes of LNG are required to produce the same energy content as conventional fuel oils • LNG storage tanks are usually located on outer surfaces on the deck • It does not singularly meet IMO's carbon reduction strategy	• There are two different types of engines: low pressure and high pressure • Otto and diesel processes can be applied • It can reduce operational costs • Flexible fuel changeovers can be made between fuel oil and LNG • The supply chain for bunkering is under development • Cost-effective clean fuel	• Methane slip • Boil-off
Methanol	• It has a lower carbon ratio than conventional fuels • It can reduce CO_2 emissions by approx. 25% • It provides an effective reduction in SOx and PM emissions due to the clean-burning properties of methanol • It has been approved by The IMO Maritime Safety Committee that it can be used as fuel on ships • It is easier to store and use on ships than other alternative fuels • It is liquid at ambient temperature	• It has a lower energy density than fuel oils • Larger volumes of methanol are required to produce the same energy content as conventional fuel oils • Exhaust treatment systems may be required to achieve IMO Tier III emission levels • It does not meet the IMO carbon reduction strategy singularly • It may be flammable when compared to others because its flammable range in the air is between 6% and 36.5% • Special fire extinguishing equipment should be used	• It can be used on ships by making minor modifications to existing systems • It has been used around the world for many years. Existing infrastructure can be modified to supply ports and ships • It can be easily stored with small arrangements to be made in the existing fuel tanks on the ships • It is currently considered the 4th most common marine fuel	• It is toxic and poisonous • Overexposure can cause death • It is corrosive to certain materials • Methanol vapor is heavier than air. For this reason, it may accumulate at points such as tank bottoms and pose a risk to seafarers

(continued)

(continued)

Types of fuels	Strengths	Weaknesses	Opportunities	Threats
Ammonia	• It proposes a zero-carbon emissions composition for the maritime industry • It meets IMO's initial GHG emission strategy • It can be stored as a liquid on ships at 20 °C and 8.6 bar (relatively higher temperature and lower pressure) • It has lower flammability when compared to others because its flammable range in air is between 15.15% and 27.35%	• Due to its structure, it requires a high proportion of pilot fuel for ignition • It has a lower energy density than fuel oils • Larger volumes of ammonia are required to produce the same energy content as petroleum-based fuels • For the safety of seafarers, exposure levels should be limited • It has poor combustion properties in internal combustion engines • SCR system can be installed to reduce NOx emissions • Fuel infrastructure for bunkering is insufficient • Fuel applications on ships are complex and have high costs compared to other systems	• The use of ammonia fuel is being developed for dual-fuel (DF) engines • It can be produced from fossil fuels using methods such as carbon capture or renewable energy	• It is considered a dangerous substance due to its toxic nature • Depending on the concentration exposed, it can irritate the eyes, lungs, and skin or be life-threatening by direct contact • The IGF code does not cover the use of NH_3 • It is not compatible with all materials due to its corrosive effect • Due to its characteristics, there is an increase in NOx emissions as a result of combustion in engines • It causes CO_2 release in global terms since the current production process is realized by HFO or coal

Sources Hansson et al. (2020), Xing et al. (2021), Wan et al. (2015), Chu et al. (2019), Gilbert et al. (2018), Alvela et al.(2018), Valera and Agarwal (2019;) ABS (2020a), Mallouppas and Yfantis (2021), Cheliotis et al. (2021), MAN Energy Solutions (2020), Ampah et al. (2021), Karatug et al. (2022), ABS (2021) ABS (2020b), Natural Resources Canada (2013), Salarkia and Golabi (2021)

References

ABS (2020a) Ammonia as marine fuel

ABS (2020b) LNG as marine fuel

ABS (2021) Methanol as marine fuel

Al-Breiki M, Bicer Y (2020) Technical assessment of liquefied natural gas, ammonia and methanol for overseas energy transport based on energy and exergy analyses. Int J Hydrogen Energy 45:34927–34937. https://doi.org/10.1016/J.IJHYDENE.2020.04.181

Al-Haidous S, Al-Breiki M, Bicer, Y, Al-Ansari T, Pecht G, Lee M, Banaitis A (2022) Evaluating LNG supply chain resilience using SWOT analysis: the case of Qatar. Energies 15:79. 15 (2021) 79. https://doi.org/10.3390/EN15010079.

Alvela M, Carroll DC, Marten I (2018) Global Gas Report 2018:2–5

Ammar NR (2019) An environmental and economic analysis of methanol fuel for a cellular container ship. Transp Res Part D Transp Environ 69:66–76. https://doi.org/10.1016/j.trd.2019.02.001

Ampah JD, Yusuf AA, Afrane S, Jin C, Liu H (2021) Reviewing two decades of cleaner alternative marine fuels: towards IMO's decarbonization of the maritime transport sector. J Clean Prod 320:128871. https://doi.org/10.1016/j.jclepro.2021.128871

Andersson K, Brynolf S, Hansson J, Grahn M (2020) Criteria and decision support for a sustainable choice of alternative marine fuels. Sustainability 12. https://doi.org/10.3390/su12093623

Balasubramanian D, Inbanaathan PV, Gugulothu SK, Noga M (2021) Characterization of single-cylinder di diesel engine fueled with waste cooking oil biofuel/diesel blends. In: Singh AP, Kumar D, Agarwal AK (eds) Altern. fuels adv. combust. tech. as sustain. solut. intern. combust. Engines. Springer Singapore, Singapore, pp 173–196. https://doi.org/10.1007/978-981-16-1513-9_8

Balcombe P, Brierley J, Lewis C, Skatvedt L, Speirs J, Hawkes A, Staffell I (2019) How to decarbonise international shipping: Options for fuels, technologies and policies. Energy Convers Manag 182:72–88. https://doi.org/10.1016/j.enconman.2018.12.080

Belgiorno G, Di Blasio G, Shamun S, Beatrice C, Tunestål P, Tunér M (2018) Performance and emissions of diesel-gasoline-ethanol blends in a light duty compression ignition engine. Fuel 217:78–90. https://doi.org/10.1016/j.fuel.2017.12.090

Bilgili L (2021) Comparative assessment of alternative marine fuels in life cycle perspective. Renew Sustain Energy Rev 144:110985. https://doi.org/10.1016/J.RSER.2021.110985

Di Blasio G, Belgiorno G, Beatrice C (2017) Effects on performances, emissions and particle size distributions of a dual fuel (methane-diesel) light-duty engine varying the compression ratio. Appl Energy 204:726–740. https://doi.org/10.1016/j.apenergy.2017.07.103

Natural Resources Canada (2013) Liquefied natural gas: properties and reliability, Gov. Canada, pp 1–3

Ceylan BO, Akyuz E, Arslanoğlu Y (2022a) Modified quantitative systems theoretic accident model and processes (STAMP) analysis: a catastrophic ship engine failure case. Ocean Eng 253:111187. https://doi.org/10.1016/J.OCEANENG.2022.111187

Ceylan BO, Karatuğ Ç, Arslanoğlu Y (2022b) A novel methodology for the use of engine simulators as a tool in academic studies. J Mar Sci Technol. https://doi.org/10.1007/s00773-022-00902-9

Cheliotis M, Boulougouris E, Trivyza NL, Theotokatos G, Livanos G, Mantalos G, Stubos A, Stamatakis E, Venetsanos A (2021) Review on the safe use of ammonia fuel cells in the maritime industry. Energies 14:1–20. https://doi.org/10.3390/en14113023

Chrysafis KA, Theotokas IN, Lagoudis IN (2022) Managing fuel price variability for ship operations through contracts using fuzzy TOPSIS. Res Transp Bus Manag 43:100778. https://doi.org/10.1016/J.RTBM.2021.100778

Chu Van T, Ramirez J, Rainey T, Ristovski Z, Brown RJ (2019) Global impacts of recent IMO regulations on marine fuel oil refining processes and ship emissions. Transp Res Part D Transp Environ 70:123–134. https://doi.org/10.1016/j.trd.2019.04.001

Deng J, Wang X, Wei Z, Wang L, Wang C, Chen Z (2021) A review of NOx and SOx emission reduction technologies for marine diesel engines and the potential evaluation of liquefied natural gas fuelled vessels. Sci Total Environ 766:144319. https://doi.org/10.1016/j.scitotenv.2020.144319

Deniz C, Zincir B (2016) Environmental and economical assessment of alternative marine fuels. J Clean Prod 113:438–449. https://doi.org/10.1016/j.jclepro.2015.11.089

Du Y, Chen Q, Quan X, Long L, Fung RYK (2011) Berth allocation considering fuel consumption and vessel emissions. Transp Res Part E Logist Transp Rev 47:1021–1037. https://doi.org/10.1016/J.TRE.2011.05.011

Efe B, Efe ÖF, Ishizaka A (2022) A model proposal to examine the effects of ships to marine pollution in terms of internal and external factors. Soft Comput 26:2121–2134. https://doi.org/10.1007/S00500-021-06626-Z/TABLES/14

Elidolu G, Akyuz E, Arslan O, Arslanoğlu Y (2022) Quantitative failure analysis for static electricity-related explosion and fire accidents on tanker vessels under fuzzy bow-tie CREAM approach. Eng Fail Anal 131:105917. https://doi.org/10.1016/j.engfailanal.2021.105917

Fraioli V, Beatrice C, Di Blasio G, Giacomo B, Marianna M (2017) Multidimensional simulations of combustion in methane-diesel dual-fuel light-duty engines. SAE Tech Pap 2017-01-05

Gilbert P, Walsh C, Traut M, Kesieme U, Pazouki K, Murphy A (2018) Assessment of full life-cycle air emissions of alternative shipping fuels. J Clean Prod 172:855–866. https://doi.org/10.1016/j.jclepro.2017.10.165

DNV GL (2019) Assessment of Selected Alternative Fuels and Technologies

Hansson J, Månsson S, Brynolf S, Grahn M (2019) Alternative marine fuels: Prospects based on multi-criteria decision analysis involving Swedish stakeholders. Biomass Bioenerg 126:159–173. https://doi.org/10.1016/J.BIOMBIOE.2019.05.008

Hansson J, Brynolf S, Fridell E, Lehtveer M (2020) The potential role of ammonia as marine fuel-based on energy systems modeling and multi-criteria decision analysis. Sustain. **12**, 3265. 12 (2020) 3265. https://doi.org/10.3390/SU12083265.

Hossain KA, Zakaria NMG, Sarkar MAR (2017) SWOT Analysis of China Shipbuilding Industry by Third Eyes. Procedia Eng. 194:241–246. https://doi.org/10.1016/J.PROENG.2017.08.141

Iannaccone T, Landucci G, Tugnoli A, Salzano E, Cozzani V (2020) Sustainability of cruise ship fuel systems: Comparison among LNG and diesel technologies. J Clean Prod 260:121069. https://doi.org/10.1016/j.jclepro.2020.121069

IMO (1997) Annex VI of Marpol 73/78: regulations for the prevention of air pollution from ships and Nox technical code, London, UK

IMO (2014) Third IMO GHG study executive summary, London, UK

IMO (2018) Initial IMO strategy on reduction of GHG emissions from ships

IMO (2020) Fourth IMO GHG study executive summary. London, UK

Inal OB, Deniz C (2020) Assessment of fuel cell types for ships: Based on multi-criteria decision analysis. J Clean Prod 265:121734. https://doi.org/10.1016/J.JCLEPRO.2020.121734

Inal OB, Zincir B, Deniz C (2022) Investigation on the decarbonization of shipping: An approach to hydrogen and ammonia. Int J Hydrogen Energy. https://doi.org/10.1016/J.IJHYDENE.2022.01.189

Karatuğ Ç, Arslanoğlu Y (2022) Importance of early fault diagnosis for marine diesel engines: a case study on efficiency management and environment. Ships Offshore Struct. 17:472–480. https://doi.org/10.1080/17445302.2020.1835077

Karatuğ Ç, Durmuşoğlu Y (2020) Design of a solar photovoltaic system for a Ro-Ro ship and estimation of performance analysis: a case study. Sol Energy 207:1259–1268. https://doi.org/10.1016/j.solener.2020.07.037

Karatug C, Arslanoglu Y, Guedes Soares C (2022) Evaluation of decarbonization strategies for existing ships. Trends Marit Technol Eng 2:45–54. https://doi.org/10.1201/9781003320289-5

Kim K, Roh G, W. Kim, K. Chun, A (2020a) Preliminary study on an alternative ship propulsion system fueled by ammonia: environmental and economic assessments. J Mar Sci Eng 8:183. https://doi.org/10.3390/JMSE8030183

Kim H, Koo KY, Joung T-H (2020b) A study on the necessity of integrated evaluation of alternative marine fuels. 4:26–31. https://doi.org/10.1080/25725084.2020.1779426

Kumar C, Rana KB, Tripathi B (2021) Combustion, performance and emission analysis of diesel-methanol fuel blend in CI engine. In: Agarwal AK, Valera H, Pexa M, Čedík J (eds) Methanol a sustain. transp. fuel CI engines. Springer Singapore, Singapore, pp 229–246. https://doi.org/10.1007/978-981-16-1280-0_9

Liu M, Li C, Koh EK, Ang Z, Lam JSL (2019) Is methanol a future marine fuel for shipping? In: J. Phys. Conf. Ser., IOP Publishing

Lunde Hermansson A, Hassellöv IM, Moldanová J, Ytreberg E (2021) Comparing emissions of polyaromatic hydrocarbons and metals from marine fuels and scrubbers. Transp Res Part D Transp Environ 97:102912. https://doi.org/10.1016/J.TRD.2021.102912

Bin Din MA, Dahalan WM, Shamsuddin SA (2022) Performance evaluation of malaysian maritime business companies during Covid-19 with TOPSIS. In: Ismail A, Dahalan WM, Öchsner A (eds) Adv. marit. technol. appl. Springer International Publishing, Cham, pp 21–31

MacFarlane DR, Cherepanov PV, Choi J, Suryanto BHR, Hodgetts RY, Bakker JM, Ferrero Vallana FM, Simonov AN (2020) A roadmap to the ammonia economy. Joule 4:1186–1205. https://doi.org/10.1016/J.JOULE.2020.04.004

Mallouppas G, Yfantis EA (2021) Decarbonization in shipping industry : a review of research. Technol Dev, Innov Propos

MAN Energy Solutions (2020) Managing methane slip

McKinlay CJ, Turnock SR, Hudson DA (2021) Route to zero emission shipping: Hydrogen, ammonia or methanol? Int J Hydrogen Energy 46:28282–28297. https://doi.org/10.1016/J.IJHYDENE.2021.06.066

Olabi AG, Wilberforce T, Sayed ET, Abo-Khalil AG, Maghrabie HM, Elsaid K, Abdelkareem MA (2022) Battery energy storage systems and SWOT (strengths, weakness, opportunities, and threats) analysis of batteries in power transmission. Energy 254:123987. https://doi.org/10.1016/J.ENERGY.2022.123987

Paulauskiene T, Bucas M, Laukinaite A (2019) Alternative fuels for marine applications: biomethanol-biodiesel-diesel blends. Fuel 248:161–167. https://doi.org/10.1016/J.FUEL.2019.03.082

Perčić M, Vladimir N, Fan A (2020) Life-cycle cost assessment of alternative marine fuels to reduce the carbon footprint in short-sea shipping: a case study of Croatia. Appl Energy 279:115848. https://doi.org/10.1016/J.APENERGY.2020.115848

Perčić M, Vladimir N, Fan A (2021) Techno-economic assessment of alternative marine fuels for inland shipping in Croatia. Renew Sustain Energy Rev 148:111363. https://doi.org/10.1016/J.RSER.2021.111363

Pucilowski M, Jangi M, Shamun S, Li C, Tuner M, Bai X-S (2017) Effect of start of injection on the combustion characteristics in a heavy-duty DICI engine running on methanol. SAE Tech Pap 2017-01–05. https://doi.org/10.4271/2017-01-0560.

Salarkia MR, Golabi S (2021) Liquefied natural gas (LNG): alternative marine fuel restriction and regulation considerations, environmental and economic. Assessment 10:44–59

Shamun S, Belgiorno G, Di Blasio G, Beatrice C, Tunér M, Tunestål P (2018) Performance and emissions of diesel-biodiesel-ethanol blends in a light duty compression ignition engine. Appl Therm Eng 145, 444–452. https://doi.org/10.1016/j.applthermaleng.2018.09.067

Stavroulakis PJ, Papadimitriou S (2017) Situation analysis forecasting: the case of European maritime clusters. Marit Policy Manag 44:779–789. https://doi.org/10.1080/03088839.2017.1330560

UNCTAD (2017) Review of maritime transport 2017. United Nations,. https://doi.org/10.18356/e9e3b605-en

Valera AK, Agarwal H (2019) Methanol as an alternative fuel for diesel engines. Methanol Altern Fuel Econ, 9–33. https://doi.org/10.1007/978-981-13-3287-6_2

Verhelst S, Turner JW, Sileghem L, Vancoillie J (2019) Methanol as a fuel for internal combustion engines. Prog Energy Combust Sci 70:43–88. https://doi.org/10.1016/J.PECS.2018.10.001

Wan C, Yan X, Zhang D, Shi J, Fu S, Ng AKY (2015) Emerging LNG-fueled ships in the Chinese shipping industry: a hybrid analysis on its prospects. WMU J Marit Aff 14:43–59. https://doi.org/10.1007/s13437-015-0080-6

Wang Y, Liu P, Yao Y (2022) BMW-TOPSIS: a generalized TOPSIS model based on three-way decision. Inf Sci (ny) 607:799–818. https://doi.org/10.1016/J.INS.2022.06.018

Xing H, Stuart C, Spence S, Chen H (2021) Alternative fuel options for low carbon maritime transportation: pathways to 2050. J Clean Prod 297:126651. https://doi.org/10.1016/J.JCLEPRO.2021.126651

Yang S, Pan Y, Zeng S (2022) Decision making framework based Fermatean fuzzy integrated weighted distance and TOPSIS for green low-carbon port evaluation. Eng Appl Artif Intell 114:105048. https://doi.org/10.1016/J.ENGAPPAI.2022.105048

Zhang M, Zhang D, Fu S, Kujala P, Hirdaris S (2022) A predictive analytics method for maritime traffic flow complexity estimation in inland waterways. Reliab Eng Syst Saf 220:108317. https://doi.org/10.1016/J.RESS.2021.108317

Zincir B, Shukla P, Shamun S, Tuner M, Deniz C, Johansson B (2019a) Investigation of effects of intake temperature on low load limitations of methanol partially premixed combustion. Energy Fuels 33:5695–5709. https://doi.org/10.1021/acs.energyfuels.9b00660

Zincir B, Deniz C, Tunér M (2019b) Investigation of environmental, operational and economic performance of methanol partially premixed combustion at slow speed operation of a marine engine. J Clean Prod 235:1006–1019. https://doi.org/10.1016/j.jclepro.2019.07.044

Zincir B (2020) A short review of ammonia as an alternative marine fuel for decarbonised maritime transportation. In: Int. Conf. Energy, Environ. Storage Energy, pp 373–380

Part III
Innovative Technologies

Chapter 6
Electrification and Hybridization of Ferries: State of The Art and Case Study

Omer Berkehan Inal, Jean-Frédéric Charpentier, and Cengiz Deniz

Abstract Zero-emission maritime transportation is the ultimate goal of the shipping industry. In this aim, new regulations lead to technical constraints which lead the naval industry toward efficient and environmentally friendly power and propulsion systems. Especially, minimizing the impacts of the ships operating in the coastal, harbor, or urban areas is a key feature considering environmental and human health concerns. Shortly, ferries or passenger ships operating in these areas will face reaching zero-emission constraints. Using electricity as a main vector of energy is one of the most promising ways to reach these goals. In particular, full electric or hybrid solution using multi-source energy systems appears to be a very relevant solution. This chapter is devoted to the study of the electrification and hybridization of ferries and passenger ships operating in coastal and urban areas. The first part of the chapter aims to review the fully electric and hybrid propulsion topologies and power management strategies that can be used for ferries and passenger ships operating in coastal or urban areas. Several relevant examples of vessels using this kind of technology are presented. Particular focus will be on full electric solutions based on batteries/fuel cells/supercapacitors as main energy sources and possible energy management strategies. In the second part of the chapter, a case study will be presented based on the specifications and mission profile of the Istanbul Ferries crosses the Bosporus straight. Several possible technical solutions are studied by giving mathematical models of the equipment respecting the state of the art in the first part.

Keywords Hybrid propulsion · Electric ships · Hybrid system modelling · Emission reduction

O. B. Inal (✉) · C. Deniz
Maritime Faculty, Marine Engineering Department, Istanbul Technical University, Istanbul, Turkey
e-mail: inalo@itu.edu.tr

J.-F. Charpentier
Research Institute of the French Naval Academy (IRENav), CC 600, 29240 Brest, France

B. Zincir et al. (eds.), *Decarbonization of Maritime Transport*, Energy, Environment, and Sustainability, https://doi.org/10.1007/978-981-99-1677-1_6

Nomenclature

CF	Carbon Factor
DG	Diesel Generator
EEDI	Energy Efficiency Design Index
EMS	Energy Management System
ESS	Energy Storage System
FC	Fuel Cell
GHG	Greenhouse Gas
ICE	Internal Combustion Engine
IMO	International Maritime Organization
PEMFC	Proton Exchange Membrane Fuel Cell
$P_{BAT,max}$	Maximum Battery Power
$P_{BAT,min}$	Minimum Battery Power
$P_{BAT,opt}$	Optimal Battery Power
$P_{DG,max}$	Maximum Diesel Generator Power
$P_{DG,min}$	Minimum Diesel Generator Power
$P_{DG,opt}$	Optimal Diesel Generator Power
$P_{FC,max}$	Maximum Fuel Cell Power
$P_{FC,opt}$	Optimal Fuel Cell Power
P_T	Total Power
SEEMP	Ship Energy Efficiency Management Plan
SFC	Specific Fuel Consumption
SOC	State of Charge

6.1 Introduction

Transportation is one of the vital elements of world trade. In 2017, the United States Energy Information Administration (EIA) published a study on the energy consumption of the transportation industry, according to the data, the transportation sector consumes 110 quadrillions BTU in 2015, and it is estimated that it will rise to 140 quadrillions BTU in 2040. By the way, maritime transportation is responsible for 90% of global transportation with 90% of outer and 40% of inner freight (Deniz and Zincir 2016; Fan et al. 2018). This high amount of transportation requires an energy source and this causes a huge amount of fuel consumption. The shipping industry mainly uses high sulfur content fossil fuels, and International Maritime Organization (IMO) states that ships approximately consume 300 million tons of fossil fuels annually. This consumption causes to emit greenhouse gases such as NOx, PM, VOC, and SOx emissions. Maritime transportation corresponds to 3.1% when the total CO_2 emissions worldwide are considered (Balcombe et al. 2019). In this connection, greenhouse gas emissions are important subjects for the shipping industry in the meaning of long-term commerce and sustainability. The IMO works on radically eliminating ship-sourced emissions by updating and/or establishing new rules and regulations, such as Energy Efficiency Design Index (EEDI), Ship Energy Efficiency Management Plan (SEEMP) in the International Convention for the Prevention of Pollution from Ships (MARPOL) Annex VI (IMO 2015). While the main goal of international shipping is to minimize emissions, inland shipping is also highlighting zero-emission transportation for environmental and public health concerns. However, the mentioned emissions are inevitable, while the fossil fuels like marine diesel oil and heavy fuel oil are still used in the shipping industry (Inal and Deniz 2021). For this reason, more effective solutions or alternative power sources should be found to replace internal combustion engines in the maritime industry. Electric and hybrid propulsion is one of the possible solutions in the way of decarbonized shipping. The United Nations state that today 55% of the world's population lives in cities, and this number is expected that to increase to 68% by 2050 (UNCTAD 2020). Therefore, the emissions of the passenger ships and ferries operating in urban areas gain importance. In this perspective, to reduce ship-sourced emissions, minimization of fossil fuel consumption is the key strategy. At this point, hybrid energy sources are prominent in ships like the automotive industry.

Hybrid propulsion can decrease the fuel consumption of the marine diesel engine by supplying additional power to the system (Bennabi et al. 2017; Dedes et al. 2016; Inal et al. 2022a, b). A classical hybrid propulsion system consists of several different energy storages such as batteries, supercapacitors, and power generation equipment such as diesel engines, and fuel cells. Since the working characteristics of these types of equipment are different, a proper analysis should be carried out according to the operation profile of the ship to find the optimal hybrid design and power management strategy. Sections II and III of this chapter include deeper information about the mentioned equipment and their selection strategies.

This chapter explains the use of hybrid and electric propulsion systems which can be a solution for decarbonized maritime transportation. The chapter contains a description of hybrid propulsion topologies, power management strategies, possible energy sources, and a case study, respectively. Lastly, the chapter's conclusion discusses the role of hybrid and electric propulsion in maritime transportation.

6.2 Hybrid Systems and Power Management Strategies

Hybrid systems mainly consist of two different systems, firstly, energy storage systems (ESS) like batteries, supercapacitors, and flywheels. Secondly, power generation units like fuel cells, marine diesel engines, solar panels, and wind turbines. The components can have various connection topologies (serial or parallel) according to ship architecture and operation profile. On the other hand, power management strategies are another important factor for an effective hybrid system. There are mainly two approaches; rule-based and optimized energy management. This section gives detailed information about hybrid system components, topologies, and power management strategies.

6.2.1 Hybrid System Components

Hybrid system components can be mainly divided into two: energy storage and power generation. On the energy storage side, batteries, and supercapacitors are reviewed. For power generation, marine diesel engines, and fuel cells are the main energy sources.

6.2.1.1 Batteries

Battery technology has been improved significantly in recent years by adopting electrical energy in mobile applications like the automotive industry. Batteries are the major energy storage technology owing to their technological maturity, lower cost, and higher energy density compared to other energy storage devices. Among various types, Li-ion batteries are the major dominator of the sector thanks to their relatively higher lifetime and considerably lower cost compared to other types. Table 6.1 gives the comparison of the characteristics of the popular battery types.

Batteries are mostly popular in the shipping industry by offering great hybridization solutions. More specifically, power density, lifetime, cost, state of charge, and size are the main constraints for batteries to expand their market share for maritime transportation. For this purpose, there are many kinds of research and analyses on the battery-diesel generator and battery-fuel cell hybrid powering systems besides the fully electric ships. For instance, a fuel cell–battery hybrid configuration is analyzed for a coastal ship from many perspectives such as cost, lifetime, power, and energy density. The optimization of the configurations has been carried out to find the most effective solution (Wu and Bucknall 2020). Another case study on a dry bulk ship shows that battery–diesel hybrid configuration has the potential to reduce greenhouse gas emissions by 14% by reducing fuel consumption (Dedes et al. 2012). Batteries are not the only hybrid power element but also have an important role for fully electric ships. For example, the world's first all-electric car ferry, MF Ampere is an important project considered a milestone in electric ship technology and has been working since 2015 in Norway (Karimi et al. 2020).

Table 6.1 Characteristics of battery types (Inal et al. 2022a; Aneke and Wang 2016; Ma et al. 2020; Nuchturee et al. 2018)

Battery type	Energy density (kWh/kg)	Efficiency (%)	Lifetime (Cycle)
Lead-acid	$30–50 \times 10^{-3}$	70–90	500–1000
Nickel–cadmium	$50–75 \times 10^{-3}$	60–65	2000–2500
Nickel metal hydride	$60–100 \times 10^{-3}$	65–90	750
Li-ion	$100–200 \times 10^{-3}$	85–90	600–2000

6.2.1.2 Supercapacitors

Supercapacitors are in demand thanks to their fast charging and discharging abilities although they have a limited energy density. Supercapacitors have higher capacitance, power density, and cycling capacity than conventional types. Their usages are relevant when the operation profile of the application requires a high load transition to buffer the system power requirements (Zhou et al. 2013). The operation characteristics of the ship are vital for the selection of the energy storage device. If the system requires the ability of high load transition in a short time, supercapacitors can be beneficial, however, if the system requires high energy density with having stabilized power supply, batteries seem more advantageous. Generally, supercapacitors are in use together with batteries to enhance the system's power supply capacity. For example, for different operation profiles of an offshore supply vessel, a battery-supercapacitor hybrid system reduces fuel consumption and greenhouse gas emissions (Nebb et al. 2012).

6.2.1.3 Diesel Engines

Marine diesel engines are the conventional, most experienced, and most common way to generate power for ships. Approximately, marine diesel engines operate at 48–52% efficiency by using heavy fuel oil and marine diesel oil (Dere and Deniz 2020). The experience level, reachable spare parts, and industrial developments are the greater pros of marine diesel engines. Currently, the major problem with diesel engines is the emissions due to fossil fuel usage. The ship-sourced emission limitations are the leading obstacles that will be faced by these engines. In this aim, alternative fuels and hybrid systems are under investigation. In the scope of this chapter, hybridization opportunities of the diesel engine are mainly reviewed. The cost of fuel, high-power capacity, and reliable operational conditions allows a diesel engine to be an important part of a hybrid ship propulsion system (Bassam 2017). The hybridization of the marine diesel engine with energy storage systems is the key approach to increasing the total system efficiency. The design characteristics of marine diesel engines are to work above 60–70% load to reach optimal operation points where the fuel consumption is minimized. However, partial and unsteady loads cause inefficient working conditions and an increase in fuel consumption and emissions (Dere and Deniz 2019). Therefore, the generic power management strategy with diesel engine hybridized systems is to supply the stored electric energy during the operations with load transition like maneuvering or departure by minimizing the load fluctuation of the diesel engine without reducing the load (Huang et al. 2021).

6.2.1.4 Fuel Cells

Fuel cells are electrochemical devices that convert the chemical energy of the fuel to electric energy without requiring any additional energy conversion like internal combustion engines (Barbir 2005; Inal and Deniz 2020). The direct energy conversion ability allows fuel cells to reach higher overall efficiency. Fuel cells generally use hydrogen but liquefied natural gas, diesel oil, or methanol also can be used after a reforming process. Fuel cells are promising options for the shipping industry owing to their emission-free power generation ability. The main reason for the produced clean energy is that it uses hydrogen as a fuel. In this chapter, fuel cells are considered power generation units due to similar working principles as diesel generators. They can produce power as long as the fuel is supplied and this form the major difference from the batteries.

Fuel cells can be classified according to their electrolyte types. The proton exchange membrane fuel cells are the leading fuel cell types among five others thanks to their technological maturity and wide application area (De-Troya et al. 2016). Fuel cells' power capacity is at the kW level, but there are several industrial types (high-temperature working types) that can generate power at the MW level (Biert et al. 2016). Fuel cells are modular, so the physical properties are changeable according to the system's requirements. The selection of the fuel cell type is highly dependent on the ship's physical properties and operation profile. Generally, fuel cells are more suitable for short-sea shipping like inland sea transportation. There are various research projects and analyses for different types of fuel cells in ships. For instance, the FellowSHIP project, one of the pilot projects for the fuel cell application in ships, was carried out from 2003 to 2018 to investigate the hybrid fuel cell-battery system for an offshore supply vessel (Tronstad et al. 2017). The system worked for 18,500 h without emitting any NO_x, SO_x, and PM.

6.2.2 *Hybrid System Topologies*

Hybrid systems can be configured under three different topologies. Generally, hybrid systems consist of various energy storage devices and power generation units which are described in the previous section. The difference between configurations is the energy distribution way from the energy sources to the propeller. The serial, parallel, and serial-parallel architectural combinations form the main categories. To simplify the description in this paper, electrical sources are considered as DC electric sources connected to a DC grid but the use of an AC grid is also possible.

6.2.2.1 Serial Configuration

The serial configuration is the most mature topology for ships compared to other configurations (Geertsma 2019). In this configuration, all energy sources are connected bus bar (or main grid) via converters. An example of this configuration is given in Fig. 6.1. In this example, an energy source, a fuel cell, a renewable energy source, and an internal combustion engine are connected to the main bus bar via converters. The propellers are driven with electric motors which receive the required

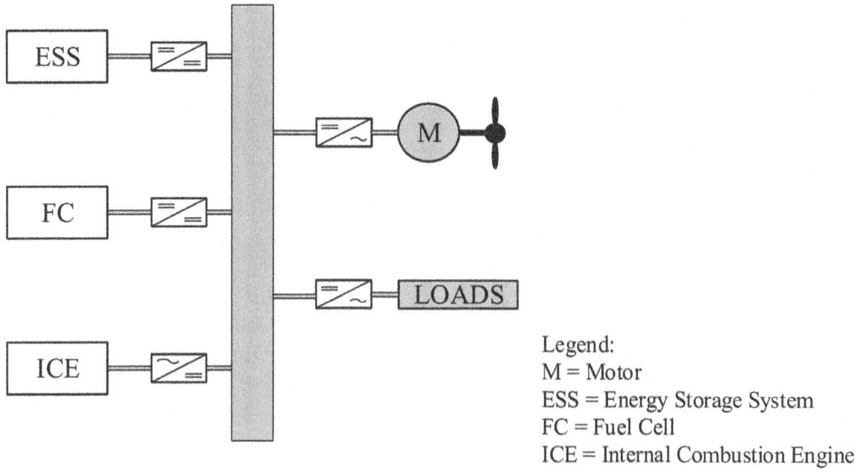

Fig. 6.1 Serial hybrid configuration (Inal et al. 2022a)

energy from the bar. In addition, other loads such as navigation equipment, HVAC, or lighting are also connected to the main grid.

The main advantage of this configuration is the capability of supplying power for various operation modes. The total electrical energy of the system is collected in the main grid and distributed to the required load. This system allows for multiple working modes with different power generation and load combinations. This topology has a relatively long operational life, and zero noise and emission are available. In contrast, multiple energy conversions and loss of energy during conversions are the major drawbacks (Nguyen et al. 2020).

6.2.2.2 Parallel Configuration

Parallel configuration differs from the serial configuration with a direct mechanical link from the diesel engine to the propellers. The electrical components of the hybrid system are similar to the serial configuration. Energy storage devices and other electric generators like fuel cells are connected to the main grid. The major advantage of this configuration is the availability of both electrical and mechanical propulsions. Also, both sizing optimization, zero-emission, and higher efficiency are allowed with this configuration. The weaknesses of the mechanical propulsion like weak maneuvering capacity or diesel engine inefficiency during low loads can be eliminated with a proper power management strategy. On the other hand, the additional mechanical links, and the need for robust control for higher efficiency can be counted as their disadvantages. This configuration is shown in Fig. 6.2.

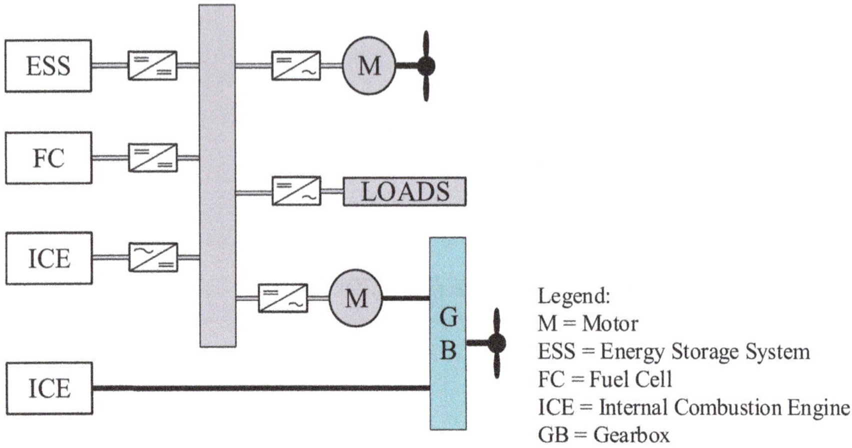

Fig. 6.2 Parallel hybrid configuration (Inal et al. 2022a)

6.2.2.3 Serial-Parallel Configuration

The hybrid (serial-parallel) configuration is the combination of both topologies. In this configuration, the internal combustion engine (generally diesel engines for ships) is connected directly to the propeller via a gearbox. Meanwhile, another gearbox connects the engine to the main electric grid. Therefore, this hybrid system allows the system to work serially and also parallelly. The architecture of the hybrid system is more complex than the other configurations, so, it requires a more detailed energy management strategy but is relatively flexible. The system contains both mechanical and electrical propulsion availability, and it can reach higher efficiency at load transitions. However, complex system architecture causes higher costs compared to other types. The generic system overview is given in Fig. 6.3.

6.2.3 Power Management Strategies

Hybrid propulsion systems have multiple energy sources and consumers. The elements of the systems have different operational characteristics and dynamic behaviors during the ship's operations. Therefore, establishing a proper energy management strategy is very important for hybrid propulsion ships to ensure safe and efficient operation. In general, costs, system performance, lifetime, and efficiency are the functions of the established energy management strategy (Garcia et al. 2010). For the same system, two different energy management strategies can have two different results during the operation. According to the aim of the system, an appropriate strategy would give the minimum fuel consumption to minimize the emission at the system's

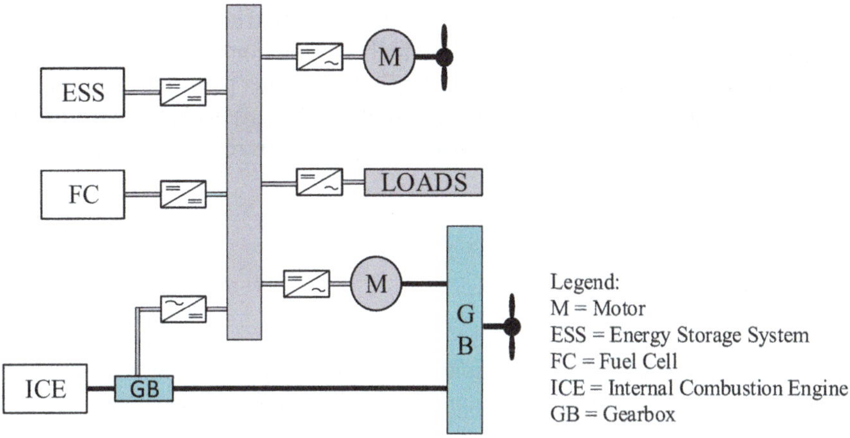

Fig. 6.3 Serial-parallel hybrid configuration (Inal et al. 2022a)

most efficient point. Hence, mainly there are two energy management strategies for hybrid ships: rule-based and optimized control strategies (Balsamo et al. 2017).

The rule-based control strategy is based on the control parameters from previous experiences. This strategy is static and robust but mostly weak for dynamic systems. The main reason for this weakness is sourced from the hardness of determining the optimal point for changes. The external factors for ships like wind or waves manipulate the expected operation and change the power need of the system dynamically. In this case, depending on the system complexity, smart control strategies like fuzzy logic gain importance for time-varying problems. For instance, in research for an all-electric ship, fuzzy logic showed better performance against a proportional-integral-based energy management strategy at the state of the charge of the batteries (Khan et al. 2017). But, instead of establishing an accurate mathematical model, fuzzy logic rules require expertise in the system behavior and mission profile.

Optimized control strategies show better performance the dynamic systems compared to rule-based energy management strategies. Global and real-time optimizations are the two main categories of the optimized control strategy. Global optimization focuses on the overall optimization during a determined time by using the mathematical equations of the system components. Many different algorithms are used in the literature to optimize the energy management of multi-energy sourced systems. Dynamic programming, genetic algorithm, artificial neural network, ant colony optimization, and particle swarm optimization are the most common techniques to perform global optimization (Ancona et al. 2018; Kumar and Fozdar 2017; Kanellos et al. 2014; Tarelko and Rudzki 2020). Real-time optimization is the second type of optimized control strategy for hybrid systems. This strategy differs from global optimization by changing according to the instantaneous conditions of the energy demand by establishing a real-time model of energy consumption. Pontryagin Minimum Principle (PMP) and equivalent consumption minimization strategy (ECMS) are the two common real-time optimization techniques. PMP is rare for ships but it has a wide application for plug-in hybrid electric vehicles. On the other hand, ECMS is used to minimize the fuel consumption of the engine with the optimum power management set points. ECMS shows better performance compared to other techniques in function of fuel consumption. For instance, by using ECMS on a hybrid ferry, a reduction in fuel consumption by 10% is reached (Geertsma et al. 2017).

6.3 Case Study

This section includes an introduction to selected case ship and mathematical models of possible hybridization equipment to analyze the pros and cons of hybrid propulsion.

6.3.1 Case Study

To assess the potential advantages of hybrid propulsion, an example ship should be taken as a case. In this chapter, ŞH-Beyoglu, a passenger ferry from Istanbul that operates at the Istanbul Strait between European and Asiatic parts of the city, has been selected. The case study ship parameters are given in Table 6.2.

The ship has a diesel-electric powering system for both propulsion and electrical loads. The ship is equipped with 2 contra rotating Schottel Twinpropeller and bow thrusters for propelling and maneuvering. The electrical loads are mainly formed by air conditioning, lighting, and navigational equipment. The generated power by diesel generators is connected to the main bus bar and distributed to the required elements such as propellers, bow thrusters, or electrical loads according to the operation profile. The ship's average voyage is 24 min and there are 65 voyages in two ways per day. The case ship's power demand and propulsion behavior change during a voyage.

Table 6.2 Ship properties

Propeller blades	3
Propeller diameter (mm)	1400
Propelling system	2 × Schottel Twinpropeller
Main engine (rpm)	1500
Main engine power (kW)	4 × 450
Main engine model	Volvo Penta D16C-A MG
Ship length (m)	67.96
Ship draft (m)	2.95
Ship gross ton	747.36
Elec. motor voltage (V)	2 × 547

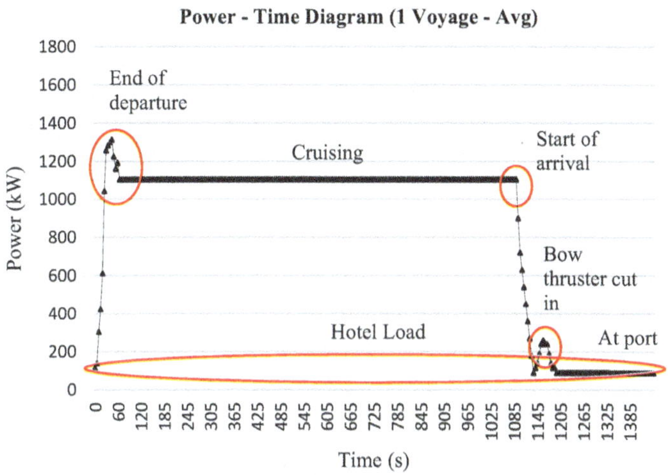

Fig. 6.4 Case ship operation profile with total consumed power (Inal et al. 2022a)

Therefore, a generic operation profile should be determined to understand the route and operation characteristics. The data according to operation are gathered physically onboard the ship. The power-related data such as energy consumption, equipment cut-in cut-off periods, and power generation fuel consumption are collected from the engine monitoring systems in the engine room. Sailing-related time-based data such as maneuvering time, bow thrusters, and propeller RPMs are gathered at the bridge by an authorized person. Multiple voyages have been completed with ship crew and average data are formed to establish a generic operation profile for the case ship for the same route. Furthermore, the operation profile is divided into four different operating conditions which are departure, steady-state, arrival, and at the port. Figure 6.4 shows the power-time diagram of the case ship operation profile.

The total route takes 24 min and the last 4 min is "at port" condition where the diesel generators work at the idle load and supply power just for electrical loads. As can be seen in the diagram, the power demand increases during the departure until the constant speed of steady-state sailing conditions is reached. This period covers the ship's speed increase and the maximum power applied to the propellers to reach the cruising speed. After the ship's speed exceeds the resistance, the demanded power drops, and the ship passes to the second phase of the operation (steady-state sailing). At the end of the steady-state sailing, the required power diminishes and the ship starts to arrive.

The peak during the arrival maneuvering is related to the bow thruster use which serves to berth at the port in a safe, efficient, and rapid way. After completion of the rope tiding, the generators work at the idle load to ensure the hotel loads.

After considering the mission profile, the following part of the paper will study a serial hybrid energy solution combining diesel generators, fuel cells, and batteries.

This solution allows the minimization of fuel consumption of DG and zero-emission modes (using FC and batteries) and fit well with the specification of the case study.

6.3.2 Model Examples

To analyze the hybrid system, the system's components should be determined, so, the most common types of equipment for a hybrid ship are listed and approaches for the mathematical modeling are explained.

6.3.2.1 Diesel Engine

Due to the case ship having a 4-stroke marine diesel engine, a 2-stroke diesel engine model is not given in this section. A dynamic 4-stroke diesel engine model can be explained with three main elements: a speed governor, an actuator, and the engine. Therefore, they can be modeled with the following transfer functions (Park 1999).

The speed governor:

$$\frac{T_s + 1}{T_1 s^2 + T_2 s + 1} \tag{6.1}$$

The actuator:

$$\frac{T_3 s + 1}{(T_4 s + 1)(T_5 s + 1)} \tag{6.2}$$

The engine:

$$e^{-T_6 s} \tag{6.3}$$

The T variables are the time constant and can be found in Park (1999). Time constants for the speed governor of the studied diesel engine T, T_1, and T_2 are 0.2, 0.0002, and 0.01 s, respectively. For the actuator, T_3, T_4, and T_5 are 0.25, 0.009, and 0.0384 s, respectively. Lastly, the time delay for the diesel engine is $T_6 = 0.024$ s. These representations are taking the time delays into account and allow the system to work dynamically. The result of the engine model would give the generated torque, and the power can be calculated at first order with the multiplication of torque with speed as follows:

$$P_{\mathrm{DG}} = T_{\mathrm{DG}} \times \omega_{\mathrm{DG}} \tag{6.4}$$

6.3.2.2 Battery

Simulink library includes a mathematical dynamic battery model that can simulate different types of batteries based on data commonly found in the datasheet manufacturers. The model makes it possible to model different types of batteries by linking an electrical model composed of a controlled voltage source in series with internal resistance to an electrochemical model, which makes it possible to link the evolution of the electrical parameters to the dynamic evolution of the state of charge of the battery.

Charging or discharging of the battery is calculated as follows:

$$E_{\text{disch}} = E_0 - K.\frac{Q}{Q - it}.i^* - K.\frac{Q}{Q - it}it + A.\exp(-B.it) \tag{6.5}$$

$$E_{\text{charge}} = E_0 - K.\frac{Q}{it + 0.1Q}.i^* - K.\frac{Q}{Q - it}it + A.\exp(-B.it) \tag{6.6}$$

where the E_{disch} and E_{charge} are the discharging and charging voltages, E_0 is the constant voltage, K is the polarization constant, Q is battery capacity and i^* is the low-frequency current, A is the exponential voltage, and lastly, B is the exponential capacity.

6.3.2.3 Fuel Cell

There are three different approaches to the modeling of fuel cells: analytical, semi-empirical, and mechanistic (Cheddie and Munroe 2005). Analytical models cover the relation between the input and output of the fuel cell according to experimental data. These types of models are quite easy to understand and do not require a long time for computation. In contrast, mechanistic ones are built respecting to chemical and physical properties of the fuel cell and are much more detailed but require a long computational time. For this reason, semi-empirical models are advantageous by combining both advantages of the other two model classes.

A generic validated fuel cell model can be found in Simulink for different fuel cell types at different power ranges.

6.3.2.4 Electric Converters

Each power source in the hybrid systems should supply different currents according to components. For this reason, converters should be used to regulate the power source voltage. There are different types of converters, to boost or reduce the voltage but most commonly the equation which describe the converter as a DC/DC tunable transformer (mean value model) can be written as follows where the V is the voltage, I is the current and η is the converter efficiency:

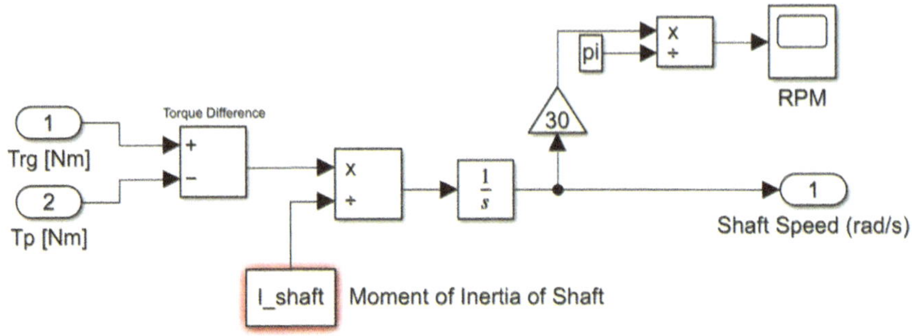

Fig. 6.5 Shaft model example

$$\frac{V_{\text{out}}}{V_{\text{in}}} = \frac{I_{\text{in}}}{I_{\text{out}}} \eta_{\text{conv}} \tag{6.7}$$

6.3.2.5 Shaft

The moment of inertia of all rotating elements should be considered while calculating the torque and power transition. Therefore, the mechanical equation of the shaft is given as follows:

$$I\frac{dy}{dt} = T_{\text{RG}} - T_{\text{P}} \tag{6.8}$$

where I is the moment of inertia of the shaft in kg m^2, and T_{RG} and T_{P} are the torque of reduction gear (applied in the propeller shaft) and the hydrodynamic torque of the propeller, respectively (these 2 torques are in Nm). The model example in the Simulink platform is shown in Fig. 6.5.

6.3.2.6 Reduction Gear

The reduction gear is used to adapt the rotational speed between the electric motor and shaft. The main parameter is the reduction ratio (k) which determines the relation between speed and torque transmission. The rotational speed and the torque are inversely proportional. Therefore, the model equation can be used as follows where the ω is rotational velocity and T is the torque:

$$\frac{T_{RG}}{T_M} = \frac{\omega_M}{\omega_{RG}} = k_{RG} \tag{6.9}$$

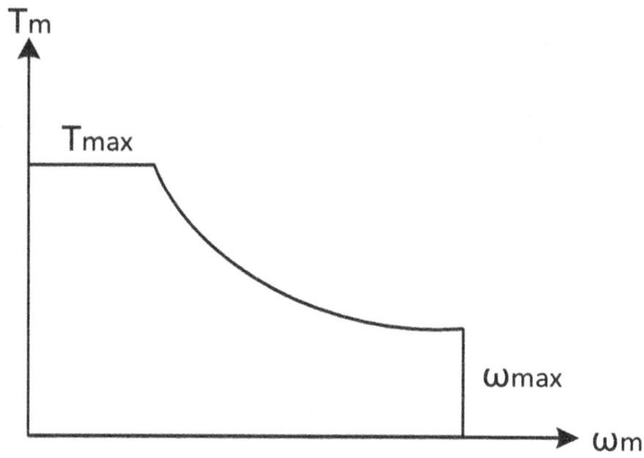

Fig. 6.6 Electric motor working limits

6.3.2.7 Electric Motor

The electric motor converts the electrical power to mechanical power for the propulsion of the ship. In our study, the power requirement of the system is known for each period of the operation, so, the modeling approach is about the torque and rotational speed-based operation limitation of the electric motor. However, the motor limitations should be considered for a more accurate calculation. Therefore, typical limitation curve of the motor is shown in Fig. 6.6, where Tm is the motor torque and ω_m is the motor rotational speed. The equation between the torque, rotational speed, and power (limitation equation) is given below:

$$T_m = \frac{P_m}{\omega_m} \tag{6.10}$$

The model example in the Simulink is shown in Fig. 6.7. There the motor parameters are the limitation couple and base speed, and a proportional integrative (PI) control is used in the model (blue rectangle). The selected maximum motor rpm is taken at 2000 RPM.

6.3.2.8 Propeller

The propeller allows to create thrust to reach the required ship speed. The propeller block is shown in Fig. 6.8. While the water speed and shaft speed are the input, propeller thrust and torque are the outputs.

The propeller calculations are using propeller geometrical data as input to calculate thrust (F_P) and torque (T_P) according to the following equations.

$$F_p = K_F \times \rho \times n_p^2 \times D_p^4 \tag{6.11}$$

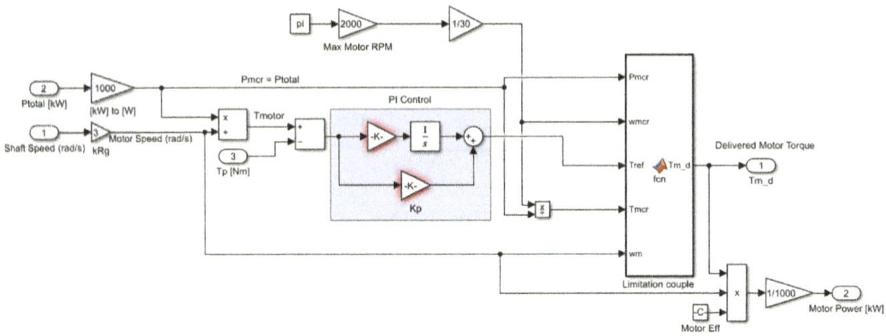

Fig. 6.7 Electric motor model example with PI control

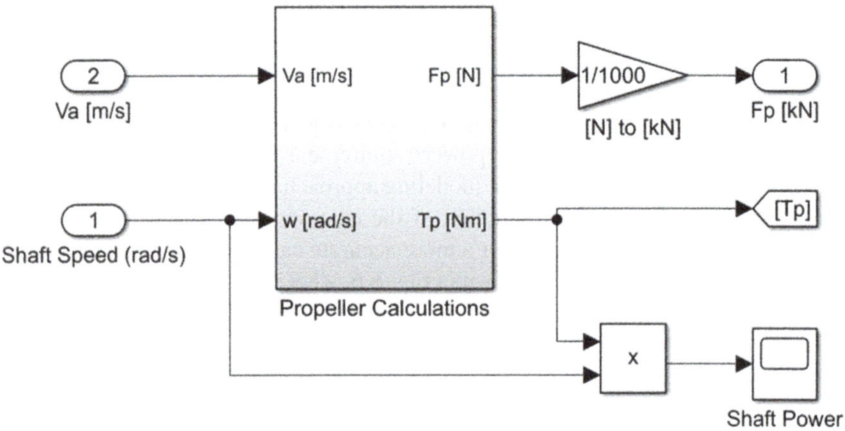

Fig. 6.8 Propulsion system example

$$T_p = K_T \times \rho \times n_p^2 \times D_p^5 \tag{6.12}$$

The non-dimensional thrust coefficient (K_F) and torque coefficient (K_T) are calculated using the following equations according to Wageningen data. These data are obtained by test of reduced scale model in basin.

$$K_F = \sum_{n=1}^{39} c_n (J)^{S_n} \left(\frac{P}{D}\right)^{t_n} \left(\frac{A_E}{A_0}\right)^{u_n} (Z)^{v_n} \tag{6.13}$$

$$K_T = \sum_{n=1}^{47} c_n (J)^{S_n} \left(\frac{P}{D}\right)^{t_n} \left(\frac{A_E}{A_0}\right)^{u_n} (Z)^{v_n} \tag{6.14}$$

The advance coefficient J can be calculated by the following equation, where the n is the rotational speed and D is the diameter of the propel

$$J = \frac{V_A}{n_p D_p} \tag{6.15}$$

ler.

Lastly, thrust force (PF) and the torque (PT) of the propeller are calculated as follows:

$$P_F = F_P \times V_A \tag{6.16}$$

$$P_T = 2\pi n_p Q_P J = \frac{V_A}{n_p D_p} \tag{6.17}$$

6.3.2.9 Hull

A proper hull model is vital for a successful hybrid ship simulation. The hull model aims to calculate the interactions between hull resistance, power, and speed of the ship. In our case, the parameters of the ship are known, therefore, ship's hydrodynamic resistance can be calculated with Holtrop-Mennen technique (Holtrop and Mennen 1982). This technique allows to calculate the total resistance and required power for a designed speed of the ship. The equation is given below:

$$R_T = R_F(1 + K_1) + R_{APP} + R_W + R_B + R_{TR} + R_A \tag{6.18}$$

where the R_F frictional resistance according to ITTC-1957, (1 + k) is the form factor, R_{APP} is appendage resistance, R_W is wave resistance, R_B is additional resistance due to bulb, R_{TR} is the additional pressure resistance due to transom immersion, and lastly, R_A is the model-ship correlation resistance. After the total resistance is calculated, the total hull model can be implemented into the general model as shown in Fig. 6.9. The resistance is dynamically changing according to ship speed and propeller thrust force.

6.3.2.10 Energy Management

Energy management strategy is extremely important for hybrid ships to ensure high efficiency with lower fuel consumption and emission. In the previous section of the chapter, most commonly used energy management strategies were reviewed. The key factor to establish a proper strategy is the operation profile and physical properties

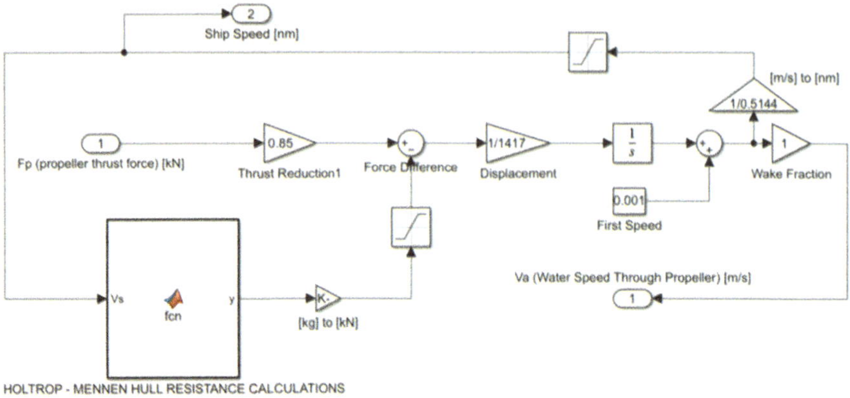

Fig. 6.9 Hull model example

of the ship. The mass and volume of the ship are important to apply the energy storage devices and their capacities. On the other side, the operation profile is used to determine the main rules and constraints to manage different power sources of the hybrid system.

In our study, according to data collected from the case ship, the average hotel load is determined as 90 kW and is considered as constant during the mission. The reached maximum power load is 1314 kW, and the average steady-state power demand is 1104 kW. Therefore, the new system should have the supply capacity of this required power. The current onboard loaded power capacity is 1800 kW with 4 diesel engines at the maximum power of 450 kW. By using the data from the engine's instruction book, the specific fuel consumption curve is calculated according to the engine load function, where the $f(x)$ is the specific fuel consumption, and x is the engine load:

$$(x) = 0.0244x^2 - 3.5228x + 317.34\left[\frac{\text{g}}{\text{kWh}}\right] \tag{6.19}$$

To find the optimal working point of the diesel engines, the specific fuel consumption curve is minimized:

$$f(x) = 190\left[\frac{g}{kWh}\right]\text{while } x = 72 \tag{6.20}$$

Therefore, the most fuel-efficient working point ($\text{P}_{\text{DG,opt}}$) of the diesel generators is at 72% load:

$$P_{DG,opt} = 450 \times \frac{72}{100} = 324kW \tag{6.21}$$

The possible diesel generator power is given below;

$$P_{DG,min}(t) < P_{DG}(t) < P_{DG,max}(t) \tag{6.22}$$

where the $P_{DG,min}$ is 40% of the maximum power ($P_{DG,max}$) (450 kW);

$$P_{DG,\min} = 450 \times \frac{40}{100} = 180\,\text{kW} \tag{6.23}$$

There are 2 diesel generators, the total power at optimal operating point can be found;

$$P_{DG,opt,total} = 324 \times 2 = 648\,\text{kW} \tag{6.24}$$

The similar approach can be followed for the fuel cell systems. Four packs of proton exchange membrane fuel cell with the maximum 100 kW power capacity which can be used optimally at 85.5% capacity. The corresponding relations for the fuel cells are given in Eqs. (3.24) to (3.26).

$$P_{FC,opt} = 85.5\,\text{kW} \tag{6.25}$$

$$P_{FC,max} = 100\,\text{kW} \tag{6.26}$$

$$0 < P_{FC}(t) < P_{FC,max}(t) \tag{6.27}$$

There are 4 fuel cell units, so the total optimal power of the fuel cells can be found as;

$$P_{FC,opt,total} = 85.5 \times 4 = 342\,\text{kW} \tag{6.28}$$

where $P_{FC,opt}$, $P_{FC,max}$, P_{FC}, and $P_{FC,opt,total}$ are the optimal power, the maximal power, the current power of one fuel cell unit, and the total optimal power to be provided by the four fuel cell units, respectively.

The state of charge of the battery is divided into three sections for establishing a proper power management system. Respecting the state of the charge, the role of the battery changes. The selected battery is a Li-ion type with an energy capacity of 200 kWh. Table 6.3 gives the main strategy for the use of batteries, respecting the state of charge the battery (SOC).

So, the power balance of the ship can be written as follows. The total power load (Fig. 6.4) power is sum of the generated and stored power from diesel generator, fuel cell and batteries. $P_{BAT,\text{dc}}$ symbolizes the discharging position of battery and $P_{BAT,\text{c}}$ is the charging position of battery. The battery's lower limit is determined as 20% of charge for a longer lifetime. For both cases, PDG varies according to the load power.

$$P_{LOAD}(t) = P_{DG} + P_{FC} {}_{-}^{+} P_{BAT} \tag{6.29}$$

Table 6.3 Battery using strategy with respect to state of charge

20–40%	40–90%	90–100%
MIN	INTER	MAX
Battery is mainly charging (if possible)	Battery is mainly discharging (if possible)	
SOCMin < SOCBAT < SOCInt	SOCInt < SOCBAT < SOCMax	

The state of the charge of the battery is the key performance indicator for the power split. So, according to min, inter, or max cases of the state of charge the battery, the rule changes. Operational priority is always to maintain the fuel cell system at its optimal operation point given by Eq. 6.27 due to zero emission compared to diesel engine. In case of emergency, the SOCmin can act as SOCinter. The summary of the charging and the discharging operations of the battery is shown in Fig. 6.10. The behavior of the battery changes according to two variables: ship's demand and SOC. The red arrows show the way where the battery SOC moves and areas are numbered respecting the instant operation and state of the charge.

There are three different states have been defined for the battery system as below:

- if $SOC < SOC_{min}$
  No 1: Battery charging @ port, arrival, and steady-state sailing.
  No 3: Neither charging nor discharging @ departure
- if $SOC_{min} < SOC < SOC_{inter}$
  No 1: Battery charging @ port, arrival, and steady-state sailing until reaching SOCinter.

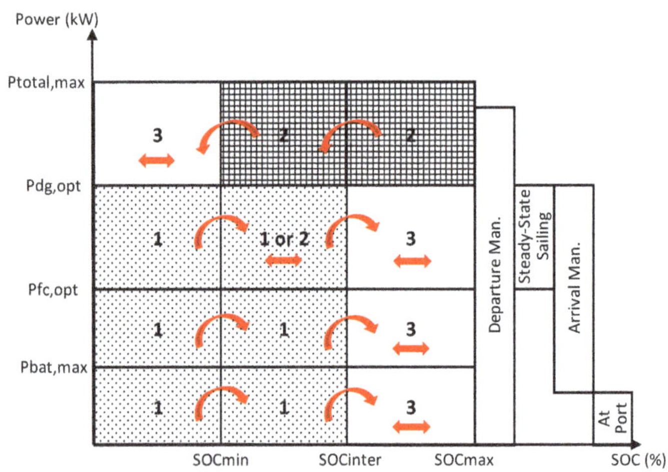

Fig. 6.10 Summary of the operation strategy

No 2: Battery discharging if SOCinter has been reached and power demand is higher than PFC, opt, total.

- if $SOC_{inter} < SOC < SOC_{max}$
  No 2: Battery discharging for power demand higher than the power which can be provided by fuel cells (battery provides a complementary power, mainly @departure and steady-state sailing.
  No 3: Neither charging nor discharging @ port, arrival, and steady-state sailing when the power demand is lower than the power which can be provided optimally by fuel cells.

The number 1 case corresponds to the areas where the battery state of charge is lower than the intermediate position, and when the power demand from the motors are lower than the sum of optimal generated powers ($P_{G,opt} = P_{DG,opt,total} + P_{FC,opt,total}$). In this case, the batteries are charged with the difference between $P_{G,opt}$ and the power demand until they reach their optimal state of charge. So, the system can be used in this state during the ports, steady-state sailing and arrivals for increasing state of charge. The second state corresponds to use of the battery as an electrical source (discharging the battery). When the battery SOC has firstly reached the preferable condition (SOCinter) and the needed power is higher than $P_{FC,total,opt,}$ the battery acts as a power source to supply an additional power to the system. In this case, fuel cell units are maintained at their optimal operating points. These cases correspond to power demand is higher than fuel cell's optimal power. In these cases, batteries can be used at desired power (according to power requirement) and DGs are used to complete the required power if needed. The third state is defined for the SOC higher the inter value, the neither charging nor discharging. However, if the SOC is lower than the minimum point, the battery won't act as an additional power source, besides, the system will not charge due to the high-power requirement of the system.

The ramp rate (R) of the power sources is the other important constraint, and the equations should be added as follows (Banaei et al. 2020):

$$\left|\frac{P_{FC}(t) - P_{FC}(t - \Delta t)}{\Delta t}\right| \leq R_{FC,max} f(t)\text{while}$$

$$R_{FC,max} = 10\,\text{kW per second} \tag{6.30}$$

$$\left|\frac{P_{BAT}(t) - P_{BAT}(t - \Delta t)}{\Delta t}\right| \leq R_{BAT,max}(t)$$

$$\text{while } R_{BAT,max} = 50\,\text{kW} \tag{6.31}$$

$$\left|\frac{P_{DG}(t) - P_{DG}(t - \Delta t)}{\Delta t}\right| \leq R_{DG,max}(t)$$

$$\text{while } R_{DG,max} = 150\,\text{kW} \tag{6.32}$$

The operation profile has been divided into four section, which are departure, steady-state sailing, arrival, and port. The analysis results show that 35% less marine diesel oil is consumed in total, and the reduction in the consumed fuel for each section of the voyage is given in Figs. 6.11 and 6.12.

To understand the environmental effect of this less fuel consumption, the harmful emissions such as CO_2, NO_x, and SO_x are calculated by using the following equations:

$$CO_2 = SFC\left[\frac{g}{kWh}\right] \times CF \tag{6.33}$$

$$SO_\chi = SFC\left[\frac{g}{kWh}\right] \times 2 \times 0.97753 \times 0.005 \tag{6.34}$$

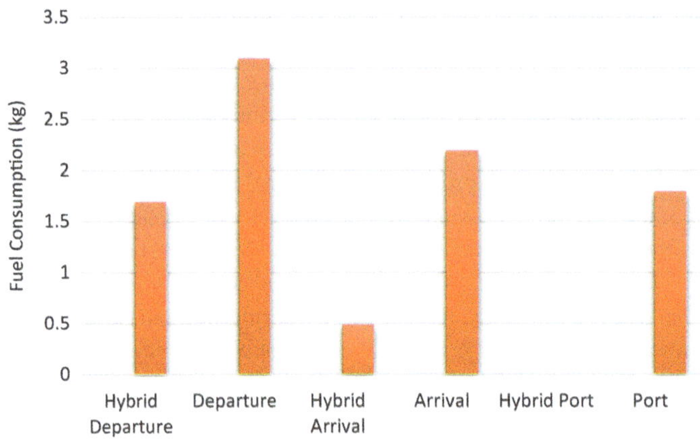

Fig. 6.11 Fuel consumptions for departure, arrival, port, and their hybrid versions for 1 voyage

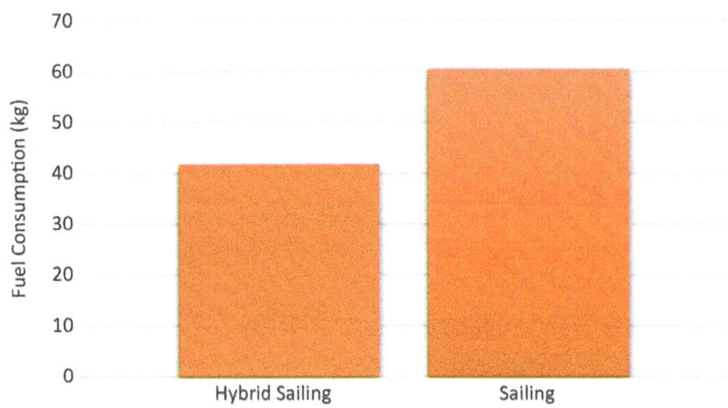

Fig. 6.12 Fuel consumptions of steady sailing and its hybrid version for 1 voyage

The specific fuel consumption (SFC) is the key factor for both emission types. In Eq. (3.32), CF symbolizes the carbon factor of the used fuel, in our case, it is 3.206, which is marine diesel oil. The Eq. (3.33) is used for SO*x* emissions (adopted from Zincir (2022)), where the 0.005 is the maximum allowable sulfur content for the fuel. The last Eq. (3.34) is for the NO*x* emissions, the engine's RPM is shown with n, and since the engines satisfy IMO Tier II limits, it can be calculated as follows:

$$NO_\chi = SFC\left[\frac{\text{g}}{\text{kWh}}\right] \times 44 \times n^{(-0.23)} \tag{6.35}$$

According to given properties, and ship's operation profile, the emission reduction potential of the proposed hybrid system is given in Figs. 6.14 and 6.15.

As can be seen in Fig. 6.13, hybrid system can reduce the fuel consumption and emissions with different ratios depending on the power demand. The decreases at the emission are 45%, 79%, and 100% for departure, arrival, and port, respectively. The reason of total emission elimination at port is due to the lower power demand than the optimal fuel cell power ($P_{FC,opt,total}$). In this position, generators and batteries are in stand-by position and hotel load is covered by the fuel cells. The results for the steady-state sailing section are given in Fig. 6.14.

As can be noticed in Fig. 6.14, the emissions during steady-state sailing have decreased by approximately 31%. The main difference is coming from the eliminating of 2 diesel generators by supplying the required power with fuel cells and batteries. During the steady-state sailing, the fossil fuel has consumed only by two diesel generators instead of four.

Total fuel consumption is about 67.9 kg per voyage but if the system were hybridized, the consumption would drop to 44.2 kg which means a reduction of about 35% per voyage. By respecting the daily and yearly data of the ferries, there

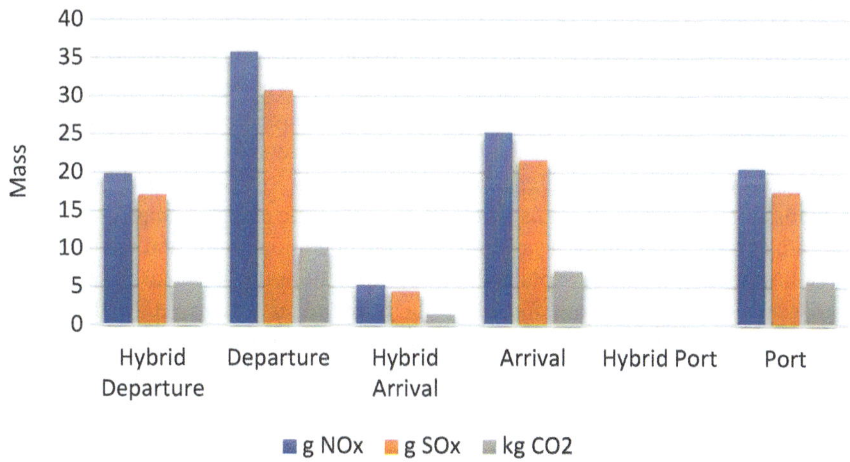

Fig. 6.13 Emissions for departure, arrival and port sections

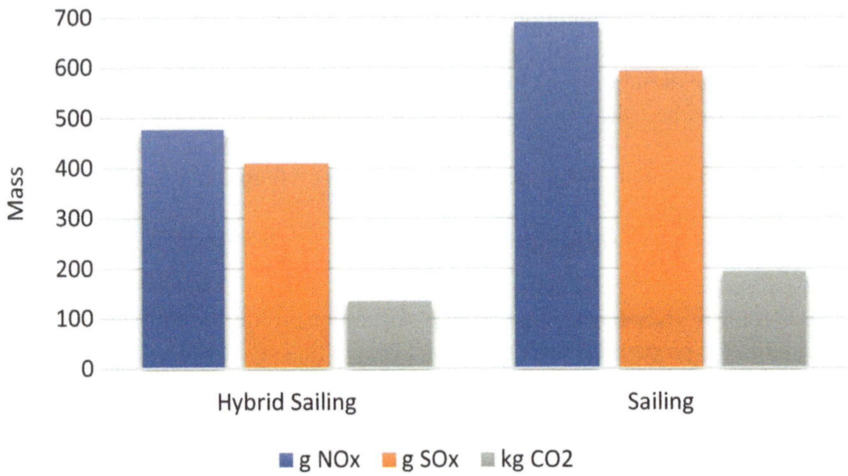

Fig. 6.14 Emissions for steady-state sailing

are 140 voyages per day, and 51,100 voyages per year (by neglecting the holidays and considering as all ferries are the same). Therefore, if the results are multiplied by 51,100, the yearly saved amount will be found 1,124 tons of diesel oil just for one ferry. The state of charge of the battery is considered as full during the analysis.

6.4 Conclusion

In today's world, greenhouse gas emissions are the main problem for ships. International Maritime Organization strongly works on emission reduction techniques and brings more strict regulations. With this aim, hybrid propulsion systems are one of the major solutions of the industry to reduce emissions. This chapter firstly introduces the hybrid ships and hybridization components with the possible topologies. Batteries, diesel generators, fuel cells, and supercapacitors are reviewed for the energy part of the systems. Serial, parallel, and serial-parallel topologies for the hybrid power systems are shown with diagrams. Then, the most popular energy management strategies are introduced. Thirdly, a case ship with the operation profile and mathematical modeling of equipment are explained. Transfer functions of a 4-stroke diesel engine, battery, and fuel cell dynamic models in the Simulink are discussed. Mathematical models of the converters and DC electric motors and propellers are presented. The mechanical equipment such as reduction gear and the shaft are represented with mathematical models. The hull resistance calculations are given with the Holtrop-Mennen technique and Simulink models are shown for the readers to enhance the clarity of the explained topic. An approach for the energy management strategy is given according to different power sources and power balance equations shown to enlighten future studies on similar topics. Lastly, reduction in fuel consumption and emissions is shown and hybrid system is compared with the conventional one. The study investigates the combined use of diesel generator, fuel cell, and batteries for the case study of an Istanbul ferry. This solution can be a very interesting choice for ferries operating into cities thanks to their zero-emission or reduction at the fuel consumption capabilities.

Acknowledgements This work was supported by the Research Fund of the Istanbul Technical University (Project ID: 43546). The authors declare that they have no known competing financial interests.

References

Ancona MA, et al. (2018) Efficiency improvement on a cruise ship: Load allocation optimization. Energy Convers Mana 164(August 2017): 42–58. https://doi.org/10.1016/j.enconman.2018.02.080.

Aneke M, Wang M (2016) Energy storage technologies and real life applications – a state of the art review. Appl Energy 179:350–377. https://doi.org/10.1016/j.apenergy.2016.06.097

Balcombe P, et al. (2019) How to decarbonise international shipping: options for fuels, technologies and policies," Energy Convers Manag 182(October 2018):72–88. https://doi.org/10.1016/j.enconman.2018.12.080.

Balsamo F, Capasso C, Miccione G, Veneri O (2017) Hybrid storage system control strategy for all-electric powered ships. Energy Procedia 126:1083–1090. https://doi.org/10.1016/j.egypro.2017.08.242

Banaei M, Ghanami F, Rafiei M, Boudjadar J, Khooban MH (2020) Energy management of hybrid diesel/battery ships in multidisciplinary emission policy areas. Energies (basel) 13(6):1–16. https://doi.org/10.3390/en13164179

Barbir F (2005) PEM fuel cells theory and practice. Elsevier

Bassam A (2017) Use of voyage simulation to investigate hybrid fuel cell systems for marine propulsion. University of Southampton

Bennabi N, Menana H, Charpentier JF, Billard J-Y (2017) Improving efficiency and emissions of small ships by the use of hybrid electrical propulsion. J Power Supply (January). https://doi.org/10.13234/j.issn.2095-2805.2017.2.001

van Biert L, Godjevac M, Visser K, Aravind Pv (2016) A review of fuel cell systems for maritime applications. J Power Sources 327(X): 345–364. https://doi.org/10.1016/j.jpowsour.2016.07.007

Cheddie D, Munroe N (2005) Review and comparison of approaches to proton exchange membrane fuel cell modeling. J Power Sources 147(1–2):72–84. https://doi.org/10.1016/J.JPOWSOUR.2005.01.003

Dedes EK, Hudson DA, Turnock SR (2012) Assessing the potential of hybrid energy technology to reduce exhaust emissions from global shipping. Energy Policy 40:204–218. https://doi.org/10.1016/j.enpol.2011.09.046

Dedes EK, Hudson DA, Turnock SR (2016) Investigation of Diesel Hybrid systems for fuel oil reduction in slow speed ocean going ships. Energy 114:444–456. https://doi.org/10.1016/j.energy.2016.07.121

Deniz C, Zincir B (2016) Environmental and economical assessment of alternative marine fuels. J Clean Prod 113(X):438–449. https://doi.org/10.1016/j.jclepro.2015.11.089

Dere C, Deniz C (2020) Effect analysis on energy efficiency enhancement of controlled cylinder liner temperatures in marine diesel engines with model based approach. Energy Convers Manag 220(March). https://doi.org/10.1016/j.enconman.2020.113015
Dere C, Deniz C (2019) Load optimization of central cooling system pumps of a container ship for the slow steaming conditions to enhance the energy efficiency. J Clean Prod 222:206–217. https://doi.org/10.1016/j.jclepro.2019.03.030
De-Troya JJ, Álvarez C, Fernández-Garrido C, Carral L (2016) Analysing the possibilities of using fuel cells in ships. Int J Hydrogen Energy 41(4):2853–2866. https://doi.org/10.1016/j.ijhydene.2015.11.145
Kumar DR, Fozdar M (2018) Optimal sizing of hybrid ship power system using variants of particle swarm optimization. In: 2017 recent developments in control, automation and power engineering, RDCAPE 2017, vol 3, pp 527–532. https://doi.org/10.1109/RDCAPE.2017.8358327
Garcia P, Fernandez LM, Garcia CA, Jurado F (2010) Energy management system of fuel-cell-battery hybrid tramway. IEEE Trans Industr Electron 57(12):4013–4023. https://doi.org/10.1109/TIE.2009.2034173
Geertsma RD, Negenborn RR, Visser K, Hopman JJ (2017) Design and control of hybrid power and propulsion systems for smart ships: a review of developments. Appl Energy 194:30–54. https://doi.org/10.1016/j.apenergy.2017.02.060
Geertsma R (2019) Autonomous control for adaptive ships: with hybrid propulsion and power generation. https://doi.org/10.4233/uuid
Holtrop J, Mennen GGJ (1982) An approximate power prediction method. Int Shipbuild Prog 29(335):166–170.https://doi.org/10.3233/ISP-1982-2933501
Huang M, He W, Incecik A, Cichon A, Królczyk G, Li Z (2021) Renewable energy storage and sustainable design of hybrid energy powered ships: a case study. J Energy Storage 43(September):103266. https://doi.org/10.1016/j.est.2021.103266
IMO (2015) Third IMO greenhouse gas study 2014. International Maritime Organization, London. https://doi.org/10.1007/s10584-013-0912-3
Inal OB, Deniz C (2020) Assessment of fuel cell types for ships: Based on multi-criteria decision analysis. J Clean Prod 265:121734. https://doi.org/10.1016/j.jclepro.2020.121734
Inal OB, Deniz C (2021) Emission analysis of LNG fuelled molten carbonate fuel cell system for a chemical tanker ship: a case study. Mar Sci Technol Bull 10:118–133. https://doi.org/10.33714/masteb.827195
Inal OB, Charpentier JF, Deniz C (2022a) Hybrid power and propulsion systems for ships: current status and future challenges. Renew Sustain Energy Rev 156. https://doi.org/10.1016/j.rser.2021.111965
Inal OB, Zincir B, Dere C (2022b) Hydrogen as maritime transportation fuel: a pathway for decarbonization. In: Agarwal AK, Valera H (eds) Greener and scalable e-fuels for decarbonization of transport. Springer Singapore, Singapore, pp 67–110. https://doi.org/10.1007/978-981-16-8344-2_4
Kanellos FD, Tsekouras GJ, Hatziargyriou ND (2014) Optimal demand-side management and power generation scheduling in an all-electric ship. IEEE Trans Sustain Energy 5(4):1166–1175. https://doi.org/10.1109/TSTE.2014.2336973
Karimi BS, Zadeh M, Suul JA (2020) Shore charging for plug-in battery- powered ships, pp 47–61
Khan MMS, Faruque O, Newaz A (2017) System for MVDC power system of all electric ship. IEEE Trans Energy Convers 32(2):798–809
Ma S, et al (2020) Fuel cell-battery hybrid systems for mobility and off-grid applications: A review. Renew Sustain Energy Rev 135(July):2021. https://doi.org/10.1016/j.rser.2020.110119
Nebb OC, Zahedi B, Lindtjorn JO, Norum L (2012) Increased fuel efficiency in ship LVDC power distribution systems. In: 2012 IEEE vehicle power and propulsion conference. VPPC 2012, pp 564–568.https://doi.org/10.1109/VPPC.2012.6422695

Nguyen HP, et al. (2020) The electric propulsion system as a green solution for management strategy of CO_2 emission in ocean shipping: a comprehensive review. Int Trans Electr Energy Syst (July): 1–29. https://doi.org/10.1002/2050-7038.12580

Nuchturee C, Li T, Xia H (2018) Energy efficiency of integrated electric propulsion for ships – a review. Renew Sustain Energy Rev 134(December):2020. https://doi.org/10.1016/j.rser.2020.110145

Park J-C (1999) Modeling and simulation of selected distributed generation sources and their assessment. West Virginia University. https://researchrepository.wvu.edu/etd/989

Tarelko W, Rudzki K (2020) Applying artificial neural networks for modelling ship speed and fuel consumption. Neural Comput Appl 32(23):17379–17395. https://doi.org/10.1007/s00521-020-05111-2

Tronstad T, Åstrand HH, Haugom GP, Langfeldt L (2017) Study on the use of fuel cells in shipping

UNCTAD (2020) Review of maritime transport 2020. United Nations Publications

van Fan Y, Perry S, Klemeš JJ, Lee CT (2018) A review on air emissions assessment: transportation. J Clean Prod 194:673–684. https://doi.org/10.1016/j.jclepro.2018.05.151

Wu P, Bucknall R (2020) Hybrid fuel cell and battery propulsion system modelling and multi-objective optimisation for a coastal ferry. Int J Hydrogen Energy 45(4):3193–3208. https://doi.org/10.1016/j.ijhydene.2019.11.152

Zhou Z, Benbouzid M, Charpentier JF, Scuiller F, Tang T (2013) A review of energy storage technologies for marine current energy systems. Renew Sustain Energy Rev 18:390–400. https://doi.org/10.1016/j.rser.2012.10.006

Zincir B (2022) Environmental and economic evaluation of ammonia as a fuel for short-sea shipping: a case study. Int J Hydrogen Energy 47:18148–18168. https://doi.org/10.1016/j.ijhydene.2022.03.281

Chapter 7
SWOT Analysis of Carbon Capture, Storage, and Transportation for Maritime Industry

Bugra Arda Zincir, Burak Zincir, and Yasin Arslanoglu

Abstract Decarbonization strategies such as renewable energy, alternative fuels, and modifications on ships play an essential role in reducing global warming, but fossil fuel consumption is inevitable for meeting the sufficient energy demand. On the other hand, CO_2 is one of the reasons for global warming, which can be mitigated by using a carbon capture system. Up to 90% CO_2 reduction can be achieved by using a carbon capture unit on a ship. Carbon capture technologies are used in power plants, cement, and the steel industry. Also, after the introduction of stricter rules and the announcement of the decarbonization target by the International Maritime Organization, some studies are started to be made on using carbon capture in maritime transportation. This chapter reviews the carbon capture technologies such as absorption, adsorption, membrane, chemical looping, cryogenic and biological, also oxy-fuel combustion, pre-combustion, and post-combustion methods. In addition, means of storage, transportation, and utilization of captured CO_2 are explained. Finally, strengths, weaknesses, opportunities, and threat analyses are done to investigate the benefits and drawbacks of carbon capture systems on vessels. Although high energy demand, storage, and transportation of the captured CO_2 are the limiting factors of adopting a carbon capture system on a ship, its high CO_2 capture rate makes carbon capture technologies a promising option to meet the recent and upcoming regulations in maritime transportation.

Keywords Carbon capture · Decarbonization · Zero-carbon · Greenhouse gases · Maritime transportation · SWOT analysis

B. A. Zincir (✉)
Maritime Vocational School, Galatasaray University, Istanbul, Türkiye
e-mail: bazincir@gsu.edu.tr

B. A. Zincir · B. Zincir · Y. Arslanoglu
Maritime Faculty, Istanbul Technical University, Istanbul, Türkiye

B. Zincir et al. (eds.), *Decarbonization of Maritime Transport*, Energy, Environment, and Sustainability, https://doi.org/10.1007/978-981-99-1677-1_7

Nomenclature

CCC	Cryogenic carbon capture
CCS	Carbon capture system
CII	Carbon Intensity Indicator
CLC	Chemical looping combustion
DCS	Data collection system
DEA	Diethanolamine
DGA	Diglycolamine
EEDI	Energy Efficiency Design Index
EEOI	Energy Efficiency Operational Indicator
EFTA	European Free Trade Association
EEXI	Energy Efficiency Existing Ship Index
EMSA	European Maritime Safety Agency
GT	Gross tonnage
IAPP	International Air Pollution Prevention
ICS	International Chamber of Shipping
IMO	International Maritime Organization
LNG	Liquefied natural gas
MAE	Methylamino ethanol
MDEA	Methyl di-ethanolamine
MEA	Mono-ethanolamine
MEPC	Marine Environment Protection Committee
MRV	Monitoring Reporting and Verification
TEA	Tri-ethanolamine
SCR	Selective catalytic reduction
SEEMP	Ship Energy Efficiency Management Plan
SWOT	Strengths, weaknesses, opportunities, threats

7.1 Introduction

The need for global trade increases with the growing population of the world. 90% of the current trade is carried out by maritime transportation (Deniz and Zincir 2016). According to a report prepared in 2021 by United Nations Conference on Trade and Development (UNCTAD), 3.8% of the maritime trade capacity is decreased due to the COVID-19 pandemic (United Nations Conference on Trade and Development (UNCTAD) 2021). However, in 2021, it is expected to increase by 4.3%. Also, in the same report, it is stated that the trade capacity by shipping will increase by 2.4% for the 2022–2026 period.

Increasing world trade causes higher fuel consumption to meet the energy demand. However, conventional fuels in the shipping industry are fossil fuels, which leads to high CO_2 generation and climate change. Even though measures have been taken to limit the global temperature increase by 2 °C in the Paris Agreement, maritime transportation is excluded. Thus, International Maritime Organization took some actions to mitigate ship-based emissions. In April 2018, Marine Environment Protection Committee (MEPC 72) was held to announce short, mid, and long-term strategies to reduce CO_2 emissions released by the ships. The strategy aims to decrease CO_2 emissions by 40% by 2030 and 70% by 2050 compared to 2008 levels (The International Council on Clean Transportation (ICCT) 2018). Details of the IMO greenhouse gas strategy are listed in Table 7.1.

In 2011, MEPC 62 was held to announce Energy Efficiency Design Index (EEDI) for ships above 400 gross tonnages (GT) (ClassNK 2022a). After two years, on 1 January 2013, the regulation entered into force to limit the CO_2 production by the new building ships. While calculating the EEDI fuel consumption, ship speed, cargo capacity, the carbon content of the combusted fuel, and the type of the ship play a role (MAN 2014). After the adoption of EEDI regulation, two types of EEDI are announced, which are required and attained. Required EEDI is the maximum grams of CO_2 per ship's capacity-mile limit for the vessel; it is calculated for each vessel. Required EEDI limits change in every phase, and every phase changes every five years, which can be seen in Table 7.2 (Zincir 2019). Attained EEDI is the actual grams of CO_2 per ship's capacity-mile produced by the ship, and the smaller it is,

Table 7.1 IMO greenhouse gas strategy (The International Council on Clean Transportation (ICCT) 2018)

Term	Year	Strategy	Target	Status
Short	2018–2023	New EEDI phases	New vessels	−10% in 2015 −20% in 2020 −30% in 2030
Short	2018–2023	Operational efficiency measures	In-service vessels	SEEMP planning required
Short	2018–2023	Improvement of the existing fleet program	In-service vessels	–
Short	2018–2023	Speed reduction	In-service vessels	–
Short	2018–2023	Measures to address VOC and methane emissions	Engine and fugitive emission	–
Mid	2023–2030	Alternative fuels implementation program	Fuels/new and in-service vessels	–
Mid	2023–2030	Further operational efficiency measures	In-service vessels	SEEMP planning required
Mid	2023–2030	Market-based measures	In-service vessels/fuels	–
Long	2030 +	Zero-carbon	Fuels/new and in-service vessels	–

Table 7.2 EEDI phases and reduction amounts (Bazari 2016).

Phase	Year	Reduction percentage
0	2013–2015	0
1	2015–2020	10%
2	2020–2025	15–20%
3	After 2025	30%

the more efficient the ship design (IMO 2022). Attained EEDI can be reduced by implementing new technologies and systems that improve energy efficiency.

With the EEDI regulation, Ship Energy Efficiency Management Plan (SEEMP) has also entered into force. SEEMP is mandatory for all ships above 400 GT to improve the efficiency of the vessels cost-effectively. Similar to EEDI regulation, SEEMP has to be implemented for each vessel depending on the cargo capacity, ship type, and route. Using a waste heat recovery system, trim-draft optimization, speed optimization, hull, propeller cleaning, and weather routing are the operational measures that are included in SEEMP (Zincir 2020). Furthermore, Energy Efficiency Operational Indicator (EEOI) is announced with the SEEMP regulation. It is a voluntary monitoring tool for companies willing to improve the energy efficiency of their ships. Companies can monitor their ship's energy efficiency by tracking emitted grams of CO_2 per ton-mile.

In July 2015 Monitoring, Reporting, and Verification (MRV) rule is implemented for vessels above 5000 GT calling to any EU and European Free Trade Association (EFTA) port (Zincir 2020). The purpose of the rule is to encourage ship owners and management companies to improve the energy efficiency of the vessel. The rule requires monitoring and reporting of voyage distance, fuel consumption, and CO_2 generated from the ships annually. The monitoring phase started in January 2018, and in January 2019 reporting phase started. Each year until April 30th, related data have to be uploaded to the THETIS database run by European Maritime Safety Agency (EMSA) (Gl 2017). Back then, every ship commercially working in EU and EFTA ports had to carry the related documents approved by EMSA.

Data Collection System (DCS) is another rule adopted by IMO in MEPC.278 (70) meeting, which was in October 2016. After two years, on 1 March 2018, the rule entered into force, and on 1 January 2019 reporting stage started. It is similar to the MRV rule, but DCS is for ships above 5000 GT calling at any ports. Every year flag state verifies the data sent by the ship, and the verified documents have to be available onboard (IMO 2019). Moreover, the rule requires the addition of a management plan for vessel fuel oil consumption, which is approved by the administration or a recognized organization, to SEEMP Part II (ClassNK 2022b).

In November 2020, at the MEPC 75 meeting Energy Efficiency Existing Ship Index (EEXI) was announced. Currently, it is not released yet, but in January 2023, it is expected to be implemented for ships above 400 GT (Gl 2021). The same concept of required and attained EEDI exists for the EEXI regulation. EEXI calculations will be made for a ship one time until a modification or alteration is made to the ship. Failure to meet the EEXI limits will cause the cancellation of the International

Air Pollution Prevention Certificate (IAPP); thus the ship will not be able to work commercially. Moreover, in the same MEPC meeting Carbon Intensity Indicator (CII) is announced, and it will also be effective after January 2023. CII is going to be mandatory for all commercially working vessels larger than 5000 GT. CII calculations will be made annually concerning grams of CO_2 generated per cargo-carrying capacity-mile (DNV GLL 2021). Each year a rating ranging from A to E will be given to the ships showing their energy efficiency. For ships to receive a D rating three times in a row or an E rating for a single time, SEEMP Part III has to be updated with the corrective action, and the plan has to be verified by the administration or recognized organization (ClassNK 2022b).

This chapter aims to give insights to its readers about carbon capture system (CCS) technologies and methods. Also, storage and transportation of the captured CO_2 are covered in Sect. 7.3 and Sect. 7.4, respectively. Besides, strengths, weaknesses, opportunities, and weaknesses (SWOT) analysis is done to investigate the advantages and disadvantages of CCS usage on ships. In the last section summary of the chapter is made, and also current concerns regarding CCS and possible future studies are explained.

7.2 Carbon Capture System

A diverse amount of CO_2 reduction means are available for the shipping industry, such as waste heat recovery systems, alternative fuels, renewable energy, hull and propeller modifications, and so on. However, applicability to ships, CO_2 reduction amounts, and costs are the limitations. Consequently, a CCS becomes an effective way to mitigate emitted CO_2 emissions by ships. Inland applications, CCS is widely used in power plants, cement, and steel industries. Nowadays, CCS has raised attention to maritime transportation, hence various studies have been made. Zhou and Wang (2014) conducted the first study on CCS installation onboard a ship (Zhou and Wang 2014). After a few years, Luo and Wang modeled a solvent-based CCS for a cargo ship in 2017 (Luo and Wang 2017). Moreover, Akker (2017) researched carbon capture application on an LNG-fueled ship, and also in 2019, another study was made by Feenstra et al. on the same topic (Akker 2017; Feenstra et al. 2019). After the announcement of stricter rules by IMO, studies on CCS have increased to meet the CO_2 reduction targets (Zincir 2020; Oh et al. 2022; Font-Palma 2021; Ji et al. 2021; Malmgren et al. 2021).

Carbon capture systems are divided into three categories which are shown in Fig. 7.1. Furthermore, different technologies are present for CCS, such as solvent-based, sorbent-based, membrane, cryogenic, chemical looping, and biological carbon capture, as can be seen in Fig. 7.2.

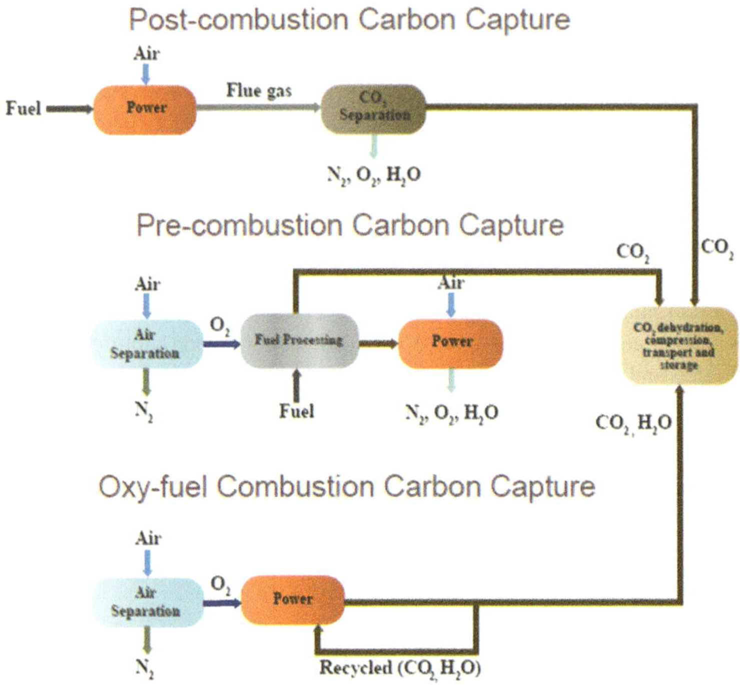

Fig. 7.1 Carbon capture methods (figure reproduced and adapted) (Repasky et al. 2014)

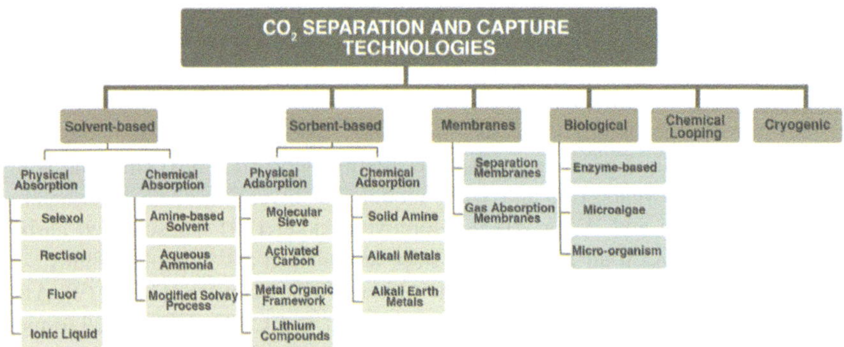

Fig. 7.2 Carbon capture technologies

7.2.1 Carbon Capture Technologies

Solvent-Based Carbon Capture

Solvent-based carbon capture technology is the most used CCS in power plants, but vessels are constantly moving platforms, hence the application of such technology requires modifications on a vessel (Luo and Wang 2017). The solvent-based carbon capture technology consists of an absorption chamber and stripper, but additional equipment such as heat exchangers, condenser, and the pump may be required depending on the used heat source.

The technology works as demonstrated in Fig. 7.3. Flue gas leaving the engine is cooled down before entering the absorption chamber. In the absorption chamber, solvent absorbs the CO_2 in the flue gas and releases the other gases into the atmosphere. Then, the solvent and the captured CO_2 mixture enter the desorption chamber, and the mixture is separated either by pressure, temperature, or electric swing. After the separation, captured CO_2 is stored in a tank while the solvent is regenerated. The carbon capture process can be achieved in two ways which are chemical and physical absorption.

The chemical absorption requires low partial pressure and temperature. Different solvents can be used in chemical absorption, such as aqueous ammonia scrubbing, amine-based solvent, and modified Solvay process (American Society of Civil Engineering (ASCE) 2014). Aqueous ammonia scrubbing, also known as chilled ammonia scrubbing, is carried out by either atomizing a solvent on flue gas or the flue gas to enter a solvent-packed bed reactor. Chemical absorption by amine-based solvent works the same as scrubbing with aqueous ammonia; the difference is the solvent and

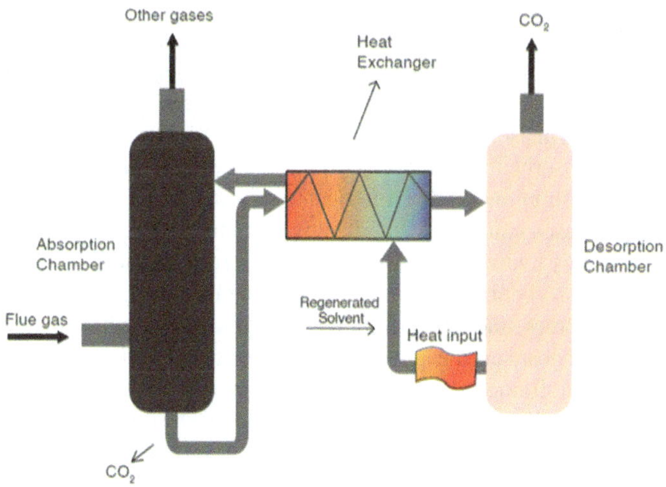

Fig. 7.3 Solvent-based carbon capture system (figure reproduced and adapted) (American Society of Civil Engineering (ASCE) 2014)

its working condition. There are different kinds of amine-based solvents, which are Mono-ethanolamine (MEA), di-ethanolamine (DEA), tri-ethanolamine (TEA), di-glycol amine (DGA), and methyl diethanolamine (MDEA) (Salvi and Jindal 2019). Due to the high carbon capture rate of MEA, it is the most preferred amine-based solvent, but its low absorption capacity makes it less desirable. Thus, in some applications, MEA is mixed with DEA or MDEA to increase carbon capture capacity (Gray et al. 2005). Apart from the amine-based and aqueous ammonia, a modified Solvay process is used to capture CO_2. The absorption process is different from the other two methods. The modified Solvay process takes place in two phases. In the first phase, a chemical reaction between CO_2 and sodium chloride occurs in an aqueous environment, while methylamino ethanol (MAE) or MEA is used as a catalyst. Then ammonium chloride and sodium bicarbonate are obtained through the reaction, and in the second phase, limestone is utilized to absorb CO_2 and release ammonia (American Society of Civil Engineering (ASCE) 2014).

The second solvent-based carbon capture method is physical absorption. During the absorption stage, a weak bond between CO_2 and the solvent occurs according to Henry's Law (Vega et al. 2017). Thus, both absorption and desorption require less energy compared to chemical absorption. Also, due to lower energy requirements, diverse amounts of solvents can be used. Physical absorption solvents are fluor, ionic liquid, rectisol, and selexol. The carbon capture process is the same with all kinds of solvents; the difference is their working temperature, pressure, and interactions with other substances. While using selexol, the optimum working temperature is 0–5 °C, and moisture in the flue gas has to be removed before. The optimum temperature for rectisol is between −1 and −38 °C, and for fluor and ionic liquids, it is the ambient temperature (American Society of Civil Engineering (ASCE) 2014).

Sorbent-Based Carbon Capture

Sorbent-based carbon capture technologies consist of an adsorber and desorber, which can be seen in Fig. 7.4. In the adsorber, a sorbent that has a porous structure is used to adsorb CO_2 to its surface by intermolecular forces (Salvi and Jindal 2019). After the adsorption phase, the sorbent and the adsorbate enter to desorber, where the regeneration of the sorbent is achieved by pressure swing, temperature swing, electrical swing, or washing methods (American Society of Civil Engineering (ASCE) 2014). Both processes can be executed by two methods. The first one is physical adsorption, it makes hydrophobic interactions, hydrogen bonds, or van der Waals bonds between the sorbent and adsorbate (Sandhyarani 2019). Temperature, partial pressure, surface forces, and pore size of the sorbent determine the efficiency of this method. Sorbents that are used in physical adsorption are activated carbon, molecular sieve, metal–organic framework, and lithium compounds.

Besides physical adsorption, chemical adsorption can be utilized to capture adsorbate. In this method, a covalent or ionic bond between the sorbent and adsorbate occurs, and the efficiency of the process is dependent on the adsorption capacity, surface area, humidity, temperature, and partial pressure of the sorbent (Ünveren et al. 2017; Kwon et al. 2011). In this method, some of the sorbents are alkali, alkali earth metals, and solid amine.

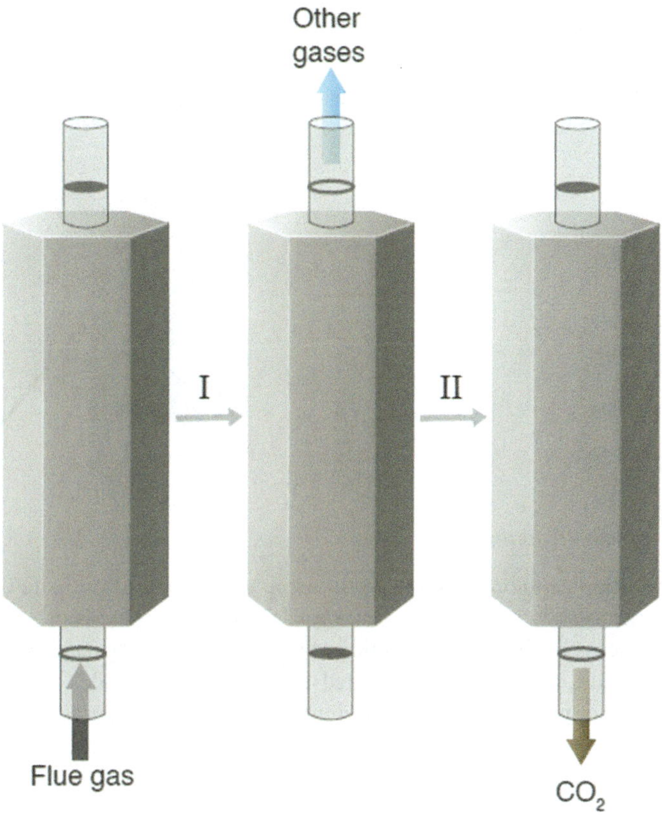

Fig. 7.4 Sorbent-based carbon capture (figure reproduced and adapted) (American Society of Civil Engineering (ASCE) 2014)

Membrane Separation

Membranes have large amounts of porous structures and act as a filter, where the permitted gases pass while the others are retained, as shown in Fig. 7.5. Selectivity and permeability are the factors affecting separation efficiency, and they differ according to the volumetric capacity and material of the membrane. Some membrane types are organic, inorganic, hybrid matrix, and facilitated transport membranes (American Society of Civil Engineering (ASCE) 2014; Abedini and Nezhadmoghadam 2010). In addition, two kinds of separation methods exist for membranes, which are gas separation and gas absorption membranes. Although they have different names, both membranes act as a filter. The only difference is gas absorption membrane has a liquid solvent in contact with the membrane thus, the filtered gas is absorbed in the solvent. Membrane separation is preferred in high-pressure stream systems such as ultrafiltration, microfiltration, forward osmosis, reverse osmosis, desalination, and medical applications (Ji and Zhao 2016).

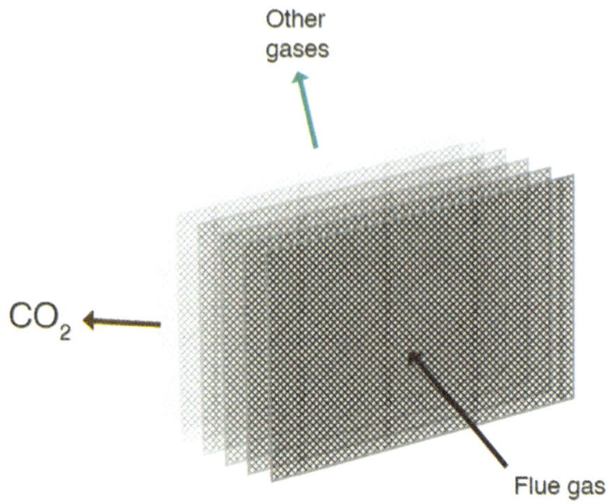

Fig. 7.5 Membrane separation

Cryogenic Carbon Capture

Cryogenic carbon capture (CCC) is a promising technology to mitigate CO_2 emissions. It works as shown in Fig. 7.6. Flue gas leaving the engine is dried before the separation process begins. Then the flue gas is cooled until CO_2 reaches its condensation temperature thus, CO_2 is separated from nitrogen dioxide, mercury, sulfur dioxide, and other gases because of different condensation temperatures (Fazlollahi et al. 2015). After that process, captured CO_2 is cooled once more until it reaches the triple point and stored in a liquid state.

Different types of CCC exist with small modifications. Stirling cooler system, cryo cell process, cryogenic distillation, anti-sublimation, controlled freezing zone, cryogenic packed bed, and external cooling loop are some of the variations, but the most mature one is cryogenic distillation (Song et al. 2019).

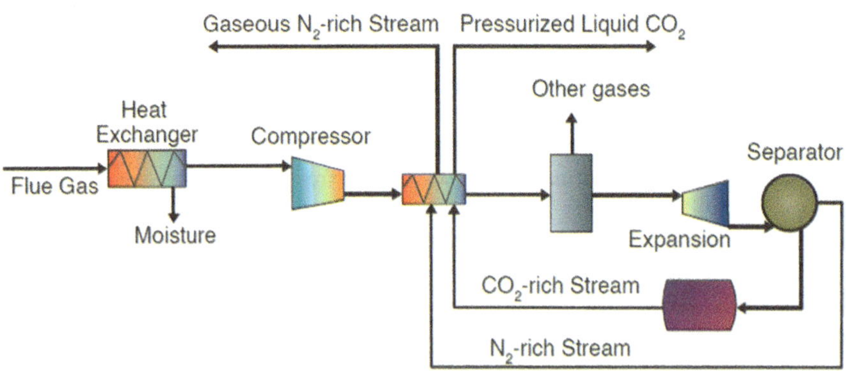

Fig. 7.6 Cryogenic carbon capture (figure reproduced and adapted) (Baxter et al. 2009)

Chemical Looping Combustion

Chemical looping combustion (CLC) is another means of capturing CO_2. The technology consists of two fluidized beds named air and fuel reactors, which can be seen in Fig. 7.7. In the air reactor metal is oxidized, and in the fuel reactor, oxygen is released by a reduction reaction hence, the oxygen is used in the combustion chamber of the engine, while the metal oxide is sent back to the air reactor (Zaman and Lee 2013). Consequently, oxygen is isolated from carbon dioxide, nitrogen, water vapor, and other gases. For the CLC system, the metal oxide material is important since it needs to be endurant for high temperatures and physical and chemical degradation (Thambimuthu et al. 2002). Some metal oxide materials used in CLC are copper dioxide, ferric oxide, manganese trioxide, and nickel oxide (American Society of Civil Engineering (ASCE) 2014).

Biological Carbon Capture

Biological carbon capture is an alternative approach to capturing CO_2. Various biological carbon capture technologies exist, and enzyme-based is one of them. Carbonic anhydrase is utilized as a catalyst to convert CO_2 into carbonic acid inside a salt solution. The process is similar to chemical absorption, but it is less energy-intensive, and low-grade heat is sufficient (Fradette et al. 2017). In addition, harnessing CO_2 to produce bio-fuel by microalgae is another biological carbon capture technology. Microalgae's simple cell structure and fast growth rate make their bio-fixation efficiency 10–50 times higher compared to terrestrial plants (Lam

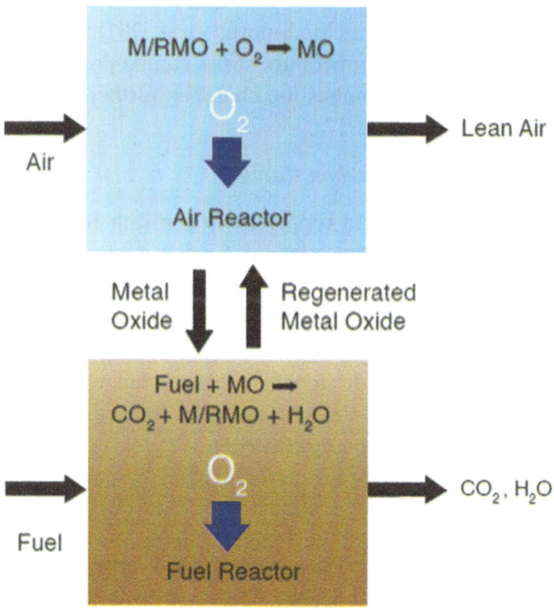

Fig. 7.7 Chemical looping combustion (figure reproduced and adapted) (Zaman and Lee 2013)

et al. 2012). The process works as, during photosynthesis, CO_2 is captured and transformed into carbohydrates, then it is used to produce bio-fuel. Besides microalgae and enzyme-based carbon capture, proteobacteria, cyanobacteria, archaea, and clostridia are used to reduce CO_2 emissions (Jajesniak et al. 2014).

7.2.2 Carbon Capture Methods

Oxy-fuel Combustion Carbon Capture

The oxy-fuel combustion carbon capture method provides oxygen for the combustion chamber and removes the other gases thus, an oxygen-enriched stream is used during combustion. Consequently, combustion products are water vapor and CO_2, but depending on the combusted fuel, sulfur oxide may appear. To obtain a pure CO_2 stream, flue gas has to be cooled down and compressed, but sulfur and ashes have to be removed before condensing water vapor (Medimurec et al. 2018).

Oxy-fuel combustion carbon capture technologies are CCC and pressure swing adsorption technologies, also Elias et al. (2018) stated that membrane separation might be used (Elias et al. 2018). A major advantage of oxy-fuel combustion carbon capture for maritime transportation is nitrogen, which is removed thus, NO_X emission is not generated (Song et al. 2019). On the other hand, this method is energy-intensive, and parasitic electric load decreases the engine efficiency (Mikulcic et al. 2019). Furthermore, high capital cost, high operational cost, stress on the materials caused by high temperature, and potential air intrusion are other drawbacks of oxy-fuel combustion capture (Medimurec et al. 2018; Pocklington and Leese 2009). Even though the high-temperature stress can be prevented by circulating a portion of the flue gas into the combustion chamber, further thermal energy loss is a disadvantage (Markewitz et al. 2012).

Pre-combustion Carbon Capture

The method aims to supply hydrogen-rich fuel to an internal combustion engine, furnace, boiler, or gas turbine. That is achieved in three steps; at first gasification, steam reforming, or partial oxidation reactions are used to produce carbon monoxide and hydrogen from fossil fuel (Medimurec et al. 2018). After that, carbon monoxide and hydrogen undergo a water–gas shift reaction in a shift converter (Mikulcic et al. 2019). CO_2 and hydrogen mixture is obtained after the reaction in the shift converter, which is also known as syngas. In the final step, CO_2 is removed, and hydrogen-rich fuel is used as the energy source.

The advantage of the method is that the product of combustion is water vapor and nitrogen; hence no harmful emissions are emitted (Salvi and Jindal 2019). Moreover, CO_2 concentration in the flue gas is higher compared to other methods. Thus required energy to remove CO_2 is lower (Ye et al. 2019). Despite the mentioned benefits, pre-combustion capture is expensive and complex due to fuel processing (Medimurec et al. 2018). Besides, the combustion temperature of hydrogen is too high compared to fuel oil, which may damage the engine. As a result, the pre-combustion capture method is preferred in hydrogen production, fertilizer production, and integrated gasification combined cycle plants (Mukherjee et al. 2019).

Post-combustion Carbon Capture

Post-combustion carbon capture removes CO_2 from the flue gas after the fossil fuel combustion. The carbon-capturing process is achieved by membrane separation, chemical and physical absorption, solid adsorption, CCC, and CLC (Mukherjee et al. 2019). Of these technologies, chemical absorption is the most mature and attractive technology for post-combustion capture (Ye et al. 2019; Aliyon et al. 2019).

An advantage of post-combustion capture is the availability of retrofitting options without modifying the combustion system. In addition, carbon capture rates can be regulated by changing the amount of flue gas entering the carbon-capturing unit. On the other hand, a disadvantage is the low CO_2 concentration in the flue gas causes high energy demand to capture it. Consequently, this method is preferred in power plants and bio-ethanol production (Mikulcic et al. 2019).

7.3 Storage of CO_2 Onboard

Captured CO_2 can be stored in solid-, liquid-, and gas states. All of them are viable methods having benefits and drawbacks. Storing in gas-state is used in pipeline transportation, but it is not a preferred solution because CO_2 in gas-state has 580 times higher volume compared to liquid form (Aspelund et al. 2006). Thus, the energy requirement for storing in gas form is too high.

Storing in solid form is another approach that has two means of doing it. The first one is to cool CO_2 until −78 °C at 1 bar, where CO_2 transforms into solid. However, that is an energy-intensive option, which requires 1146 kJ/kg-CO_2 (Akker 2017). The other option is to use the adsorbent substance to bind CO_2 to its surface. Although it sounds reasonable, the need for an extra tank to store the adsorbent substance is a disadvantage considering the loss of cargo space on a ship. Moreover, the risk of sublimation of CO_2 leads to safety risks on a ship since asphyxiation of the crew is a possibility in case of a leakage.

Storing in a liquid state is considered the most energy-efficient option for CO_2 storage. The important factor for liquid storage is temperature and pressure. For maritime transportation, storing at a triple point is recommended, which is -56.5 °C and 5.1 bar; consequently, the volume of CO_2 is minimized. However, storing at the triple point may lead to clogs in pipes due to solid formations (Akker 2017). Moreover, high amounts of CO_2 may cause stability issues on ships because of their low viscosity (Zhou and Wang 2014). Besides, water and moisture formations have to be removed before storing to prevent corrosion in pipings and storage tanks.

Considering all these, there is no commonly agreed CO_2 storage procedure. Zhou and Wang (2014) investigated both liquid- and solid-state storage (Zhou and Wang 2014). For the liquid state, the authors proposed CO_2 storage at 31 °C and 100 bar, and for solid-state, they used CaO powder as the adsorbent material. Akker (2017) stored CO_2 in a liquid state at -25 °C and 15 bar, while Feenstra et al. (2019) stored it at -16 °C and 22 bar (Akker 2017; Feenstra et al. 2019). In a study conducted by Luo and Wang (2017), a 6–16 multistage compressor is proposed to store CO_2 at 31 °C and 100 bar, but the authors also stated that the capital investment, energy requirement, and cargo space loss for the compressor is high (Luo and Wang 2017).

7.4 Transportation of Captured CO_2

Compared to ships, the carbon capture system is used more in inland facilities. Captured CO_2 is transported by pipelines in land applications since they are stationary, but this is a challenge for maritime transportation. Hence, vessels need large tanks to store captured CO_2 on board until they reach ports. For liner shipping, that issue can be fixed as the ships visit the same ports. However, voyage days for tramp transport vary, so CO_2 storage tanks should be sufficient for voyages of any duration. Either way, transportation of captured CO_2 has to be done from ports. Thus, pipelines to geographical formations, saline formations, and oil reservoirs are needed.

According to an IPCC report in 2005, CO_2 transportation from a source to a geologic site is around 1–8 USD/tCO_2 per 250 km pipeline (IPCC 2005). Moreover, before injecting CO_2, moisture and impurities such as Ar, N_2, and O_2 should be removed to improve the storage capacity and prevent corrosion (Raza et al. 2019; Wang et al. 2011). Storage of the captured CO_2 is costly in most scenarios, but it can also be beneficial. Oil enhancement of 15% can be achieved by injecting CO_2 into geological formations (Mikulcic et al. 2019). In addition, CO_2 injection can be used in oil fields to improve petrol extraction.

Another important aspect is the selection of the geologic site because temperature and pressure change according to the depth of the site. CO_2 acts as in supercritical point for geological formations having depths of more than 800 m since its temperature and pressure increase (Raza et al. 2016). During CO_2 injection, higher density ensures the efficiency and the safety of the process as the buoyancy force becomes less, and the density of the CO_2 is dependent on depth, temperature, methane, and water contamination (Raza et al. 2019).

A report issued by International Energy Agency (2009) implies that in 2050, 145 gigatonnes of space will be needed to store captured CO_2 (International Energy Agency (IEA) 2009). Besides, in the same report, saline formations had the highest storage capacity, followed by oil and gas fields is stated. Despite the high storage capacity of saline formations, injection cost is higher than the other two options.

7.5 SWOT Analysis of CCS for Ships

Strengths, weaknesses, opportunities, and threats are inevitable factors that evolve around every technology. SWOT is commonly used for qualitative analysis for strategic planning to change negative consequences to positives. By implementing SWOT analysis, a systematic thinking approach is adopted for the holistic diagnosis of new technology, product, management, or plan (Heinz 1982). Hence, SWOT analysis is used to determine the advantages and disadvantages of the utilization of CCS on ships. In Fig. 7.8, SWOT factors derived from CCS installation on ships can be seen.

7.5.1 Strengths

As can be seen from Fig. 7.8, strengths are compliance with the current CO_2 reduction regulations, meeting IMO 2030 and 2050 decarbonization targets, and cooperation with other technologies.

Current CO_2 Regulations

The MRV and DCS rules are currently in the monitoring and reporting phase. Still, sanctions are expected to be announced after 1 January 2023 with the entry into force of the EEXI and CII regulations. Moreover, EEDI and SEEMP regulations are

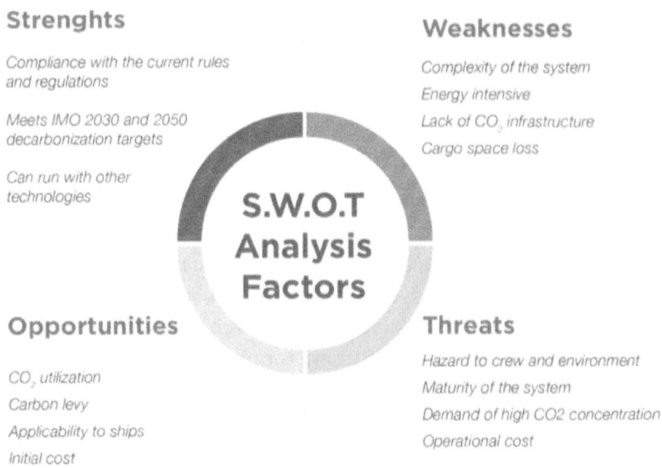

Fig. 7.8 SWOT Analysis for CCS

enforcing ship owners and managers to improve ship energy efficiency to mitigate emitted CO_2 emissions. Fortunately, carbon capture technologies provide a carbon capture rate of up to 90% thus, current and upcoming EEDI limits are met.

IMO 2030 and 2050 Decarbonization Targets

After the IMO Initial GHG Study is announced in 2018, strategies, targets for existing and new building ships, and stakeholders are introduced. Also, strategies involved in the study are listed from short to long-term measures to develop decarbonized environment. Besides, in the same study, IMO announced that the produced CO_2 needs to be under 40% by 2030 and 70% by 2050 compared to the 2008 levels. As mentioned before, by utilization of CCS on a ship, up to 90% of CO_2 emission can be captured thus, IMO's decarbonization target through 2050 is met.

Cooperation with Other Technologies

Another strength of CCS is its availability to run with other technologies that are available on a ship. CCS can be applied to ships using LNG as the cold heat source of LNG is beneficial during CO_2 storage. Also, if the proper cold energy source is supplied in a diesel-fueled ship, CCS can be run with diesel engines. Furthermore, there is no harm noted while CCS is running with other emission mitigation technologies such as selective catalytic reduction (SCR) and SO_X scrubber. A study made by Zincir (2020) demonstrates a combined model of CCS and SCR on a ship and an investigation of its performance (Zincir 2020).

7.5.2 Weaknesses

Weaknesses of CCS on ships are the system complexity, energy-intensive processes, lack of infrastructure for CO_2 transportation, and cargo space loss due to CO_2 storing and installed equipment.

System Complexity

Carbon capture technologies require additional equipment such as heat exchangers, pumps, valves, compressors, and condensers. In addition, absorber, adsorber, stripper, membranes, fuel and air reactor, and air separator may be needed for related technologies and methods. Consequently, system complexity increases because of additional pipings, pieces of machinery, and solvents. Moreover, temperature and pressure are the other important factors to handle with care for proper operation. Another problem is that the solvents or sorbents used in CCS have to be operated with care to increase carbon capture efficiency and prevention of hazards to the crew and the environment.

Energy Demand

Carbon capture technologies are energy-intensive systems mostly because of CO_2 compressing and storing. However, some technologies demand less energy than others due to the availability of cold and heat energy sources on ships. Besides

CO_2 compressing, chemical absorption, physical adsorption, membrane separation, and cryogenic and biological carbon capture technologies demand high energy during capturing process (American Society of Civil Engineering (ASCE) 2014; Ye et al. 2019; Mukherjee et al. 2019).

Infrastructure for CO_2 Transportation

Another concern for CCS on a ship is the lack of infrastructure for CO_2 transportation. Captured CO_2 can be stored onboard temporarily, but tank capacity is limited for high liquid CO_2 amounts. Thus, captured CO_2 has to be transported from ship to shore. To overcome this challenge, facilities with proper pipings and insulations need to be built. Currently, that is a setback that needs to be solved to use CCS commercially.

Loss of Cargo Space

Implementing new technology on a ship means installing new equipment, which leads to cargo space losses. Another reason for cargo space loss is caused by CO_2 storage because combusting 1 g of marine fuel produces more than 3 g of CO_2 (International Organization for Standardization (ISO) 2017). This indicates that the capacity of the tanks for CO_2 storage has to be higher than the fuel tanks. As a result, additional equipment for CCS and a tank for CO_2 storage cause less amount of freight earned from the same voyage compared to the ship without CCS.

7.5.3 Opportunities

Opportunities rise by installing carbon capture technologies on ships. These opportunities are carbon levy, applicability to ship, CO_2 utilization, and capital cost.

Carbon Levy

The carbon tax is applied to inland facilities for emitted CO_2-per ton into the atmosphere. Currently, that is not available in maritime transportation, but in September 2021, the International Chamber of Shipping (ICS), which represents 80% of the global fleet, suggested the adoption of a carbon levy for ships above 5000 GT (International Chamber of Shipping (ICS) 2021). That indicates that soon carbon tax might be applied to merchant shipping to prevent climate change caused by vessels. Moreover, IMO's decarbonization target, EEXI, and CII regulations prove that stricter measures are started to be taken in the maritime industry. Consequently, the high CO_2 mitigation potential of CCS makes it a promising technology.

Applicability to Ship

Carbon capture methods and technologies are listed and explained in Sect. 2. Unfortunately, only some are applicable to ships due to limited space, energy, and compatibility with marine engines. Oxy-fuel and pre-combustion carbon capture methods are unlikely options to be used in ships because they are hard to retrofit. However, post-combustion carbon capture is promising as, with some modifications it can be applied to ships. For carbon capture technologies, chemical absorption is predominant due to its compatibility with diesel and LNG engines. Moreover, Oh et al. (2022) show that membrane separation can be utilized on ships to capture CO_2 (Oh et al. 2022).

Utilization of CO_2

Captured CO_2 is widely used in land facilities to produce fuels, chemicals, fire extinguishers, fertilizers, and plastics. Besides, China has been using captured CO_2 in farmlands for more than 30 years (Mikulcic et al. 2019). Recently, one of Visser Shipping's vessels started using a CO_2 battery invented by Value Maritime to use captured CO_2 to grow crops and flowers in greenhouses (American Shipper 2022). However, lack of infrastructure and CO_2 transportation options are the challenges of the CO_2 utilization process to be used commercially.

Capital Cost

The initial cost of CCS is high as new equipment is needed to be implemented on vessels such as heat exchangers, pumps, compressors, and so on. However, a study conducted by Zincir (2020) compared other CO_2 mitigation technologies such as alternative fuels, renewable energy, fuel cells, and waste heat recovery system with CCS from a \$/ton-CO_2 reduction perspective, and the results demonstrate that the CCS is predominant by a quite margin (Zincir 2020).

7.5.4 Threats

As with all new technology, CCS has threats, which are risks to crew and environment, system maturity, operational cost, and high CO_2 concentration demand.

Risk to Crew and Environment

Safety and conservation of the environment in maritime transportation is a concerning factor while considering new systems. The risk of environmental pollution, flammability, toxicity, and asphyxiation plays a vital role for the crew and environment. Oxy-fuel combustion carbon capture raises the risk of fire due to high oxygen concentration, and for pre-combustion carbon capture, hydrogen gas management is an important factor since it indirectly causes global warming due to increasing the amount of water vapor, ozone, and methane (Environmental Defense Fund (EDF) 2022). Despite post-combustion carbon capture seeming innocent compared to others, solvents and sorbents that are used to capture CO_2 may harm to crew and the environment in case of a leakage. Ammonia and amine-based solvents are widely used in CCS because of their high CO_2 capture rate, but their toxicity is a disadvantage. In addition, CO_2 may cause asphyxiation of the crew because of its odorless and colorless form if there is a leakage. On the other hand, by taking the right measures, mentioned risks can be minimized.

Maturity

Research on CCS has been around for a long time, but few applications exist. Although its carbon capture rate is undeniably high compared to other CO_2 mitigation systems, it is not commercially preferred. However, CCS raised attention in the last decades as a promising solution for decarbonization, and various studies were made about it. Of the three carbon capture methods, post-combustion is the most mature one for the shipping industry and has been mentioned in studies with the chemical absorption technology multiple times (Zincir 2020; Zhou and Wang 2014; Luo and Wang 2017; Akker 2017; Feenstra et al. 2019).

High Operational Cost

The operational cost of CCS is high because carbon capture and storage are energy-intensive processes. However, CO_2 storage costs can be reduced by using an LNG-fueled engine instead of a diesel one. Moreover, by optimization of absorber and stripper chambers, Güler and Ergin (2021) managed to decrease the operational cost to 40.27 \$/ton-CO_2, which was 77.50 €/ton-CO_2 in a study conducted by Luo and Wang (2017), Güler and Ergin (2021).

CO_2 Concentration

Another threat is the demand for high CO_2 concentration in the exhaust gas. While considering oxy-fuel and pre-combustion carbon capture methods, this factor is eliminated since CO_2 concentration is higher compared to post-combustion carbon capture (Pocklington and Leese 2009; Ye et al. 2019). On the other hand, CO_2 concentration is a complicated factor for post-combustion carbon capture because amine-based solvents perform better with low CO_2 concentration, whilst ammonia becomes dominant with high CO_2 concentrations (Jilvero et al. 2014). Hence, depending on the CO_2 concentration, solvent selection should be made.

7.6 Discussion

Carbon capture systems are one of the many ways of mitigating ship-based carbon emissions. Like all the new technologies, CCS has strengths, weaknesses, opportunities, and threats. In the previous section, those advantages and disadvantages are explained. Considering the fact that stricter rules and regulations are coming to maritime transportation, CCS provides a great chance to comply with them. Moreover, after 2050, IMO aims to decarbonize the shipping industry, and current technology cannot achieve total decarbonization as a stand-alone option. Thus, combined systems stand out to achieve that, and the possibility of CCS running with the other carbon reduction technology becomes another strength of the system.

Weaknesses of the system occur because it is not a mature system for maritime transportation. Hence, the complexity of the CCS may increase the workload of the crew onboard. However, that problem is expected for every new system implemented on a ship, as personnel onboard need time to get familiar with it. Furthermore, energy-intensive processes of CCS and loss of cargo space caused by new equipment and CO_2 storage lead to less money earned from voyages. Despite those having negative impacts on earned money for the ship owner, other CO_2 mitigation technologies are also costly. Consequently, it should be expected that to compensate for the loss, freight rates need an increase to encourage decarbonization. Another weakness is the lack of infrastructure to transport captured CO_2 from ships to other places. Although that is challenging, pipelines need to be laid from ports to saline formations to overcome this issue. However, that is a costly solution to transport such a low amount of CO_2 compared to inland facilities.

An opportunity arises from CO_2 utilization. On land, captured CO_2 is harnessed to produce fuels, chemicals, fertilizer, and building material; and in the agriculture industry to grow plants. In addition, captured CO_2 can be stored in coal beds and oil reservoirs to enhance petroleum extraction. However, it has to be reminded that the amount of produced CO_2 is too low on ships compared to inland facilities. Thus, a feasibility study has to be carried out to determine the worthiness of CO_2 utilization. Adoption of a carbon levy is another opportunity for CCS since the CO_2 reduction rate is higher compared to other technologies, and paid tax will be less. The capital cost of CCS also leads to similar benefits due to its cost-effectiveness compared to the other systems. Applicability to ships is a bit contradictory factor for CCS because pre-combustion and oxy-fuel combustion carbon capture systems are not ready-to-use options for ships presently, but a post-combustion capture system is a promising option since, with small modifications on a ship, it can be implemented.

Risks affecting crew and environment are important factors considering maritime transportation. Toxicity, flammability, and asphyxiation are the risks that impact a crew. Although strict measures are taken in the shipping industry, small mistakes may lead to incidents. Thus, the used solvent and technology have to be determined with care. Besides, some solvents are harmful to species at sea. Another threat is the maturity of CCS, but it requires time to overcome that problem. CO_2 concentration seems like another threat to CCS application on ships. However, various studies were made on CCS, and the CO_2 concentration issue can be eliminated with the right solvent selection. The final threat is the operational cost of CCS. That is a similar concept to the capital cost as CO_2 reduction systems are expensive solutions. Yet its cost-effectiveness is still an important parameter to be recognized.

SWOT analysis represents the advantages and disadvantages of adopting CCS as a CO_2 reduction method. Weaknesses and threats arise because CCS is still not a commercially ready solution. However, with further research on CCS and optimizations, most of the issues can be eliminated. Moreover, implementing the right technology is a key aspect of proper and efficient operation. The positive side of the CCS is getting closer to the green ship concept. By adopting CCS, not only current requirements of the current regulations can be met, but also the upcoming decarbonization target can be achieved.

7.7 Summary

IMO aims to decarbonize maritime transportation to have a positive impact on climate change. Various systems and technologies are proposed to mitigate the emitted greenhouse gases into the atmosphere, yet current technologies are not able to achieve that, and fossil fuel consumption is increasing day by day to meet the energy demand. This chapter presents the available carbon capture technologies, storage, and transportation options. Currently, the post-combustion carbon capture method with chemical absorption technology is predominant because of its applicability to ships. While CO_2 storage, some studies were conducted on storing it in a solid state, but mostly preferred in liquid form due to cost-effectiveness. Transportation of captured CO_2 is another challenge. Presently, there is no means of transportation to CO_2 from ships to another place. However, after this issue is resolved, CO_2 can be stored in saline formations, coal beds, and geological formations.

A SWOT analysis is done to show the benefits and limitations of CCS. Although CCS provides up to 90% CO_2 mitigation, it is not commercially ready technology to be used in maritime transportation because of CO_2 storage, transportation, and energy-intensive processes. Moreover, its cost and cargo space loss are concerning factors for the ship owners and operators. On the other hand, CO_2 reduction potential, compliance with the regulations, cost-effectiveness, and cooperation with other CO_2 mitigation technologies stand out. It is concluded from the literature and SWOT analysis that the CCS may become a competitive option for decarbonization in the future. Therefore, more research is needed on CCS optimization for ships, as well as alternative ways of transporting CO_2 from ships to shore. Moreover, other carbon capture technologies than chemical absorption need to be investigated for the most efficient carbon capture system.

References

Abedini R, Nezhadmoghadam A (2010) Application of membranes in gas separation processes: its suitability and mechanisms. Pet Coal 52:69–80

Akker JT (2017) Carbon capture onboard LNG-fueled vessels, a feasibility study. M.Sc. Thesis. Department of Marine Technology, Delft University of Technology

Aliyon K, Hajinezhad A, Mehrpooya M (2019) Energy assessment of coal-fired steam power plant, carbon capture, and carbon liquefaction process chain as a whole. Energy Convers Manag 199:1–19. https://doi.org/10.1016/j.enconman.2019.111994

American Shipper (2022) Scrubber alternative can now capture carbon emissions on vessels. https://www.freightwaves.com/news/scrubber-alternative-can-now-capture-carbon-emissions-on-vessels

American Society of Civil Engineering (ASCE) (2014) Carbon Capture and Storage

Aspelund A, Molnvik MJ, Koeijer GD (2006) Ship Transport of CO_2 technical solutions and analysis of costs, energy utilization, exergy efficiency and CO_2 emissions. Chem Eng Res Des 84(9):847–855. https://doi.org/10.1205/cherd.5147

Baxter LL, Baxter A, Burt S (2009) Cryogenic CO_2 capture as cost-effective CO_2 capture process

Bazari Z (2016) IMO Train the Trainer (TTT) Course on energy efficient ship operation, module 2 - ship energy efficiency regulations and related guidelines
ClassNK (2022a) EEDI regulation. https://www.classnk.or.jp/hp/en/activities/statutory/eedi/index.html#:~:text=Amendments%20to%20MARPOL%20Annex%20VI,as%20from%201%20January%202013
ClassNK (2022b) SEEMP, IMO DCS and CII https://www.classnk.or.jp/hp/en/activities/statutory/seemp/index.html
Deniz C, Zincir B (2016) Environmental and economical assessment of alternative marine fuels. J Clean Prod 296:438–449. https://doi.org/10.1016/j.clepro.2015.11.089
DNV GL (2017) EU MRV regulation, get the details on monitoring, reporting and verifying in line with the new EU MRV regulation - the smart way to comply
DNV GL (2021) EEXI - Energy Efficiency Existing Ship Index. https://www.dnv.com/maritime/insights/topics/eexi/index.html
DNV GLL (2021) CII - Carbon Intensity Indicator https://www.dnv.com/maritime/insights/topics/CII-carbon-intensity-indicator/answers-to-frequent-questions.html
Elias RS, Wahab MIM, Fang L (2018) Retrofitting carbon capture and storage to natural gas-fired power plants: a real-options approach. J Clean Prod 192:722–734. https://doi.org/10.1016/j.jclepro.2018.05.019
Environmental Defense Fund (EDF) (2022) For hydrogen to be a climate solution, leaks must be tackled. https://www.edf.org/blog/2022/03/07/hydrogen-climate-solution-leaks-must-be-tackled
Fazlollahi F, Bown A, Ebrahimzadeh E, Baxter LL (2015) Design and analysis of the natural gas liquefaction optimization process-CCC-ES (energy storage of cryogenic carbon capture). Energy 30:1–14. https://doi.org/10.1016/j.energy.2015.05.139
Feenstra M, Monterio J, van der Akker JT, Mohammad RM, Abu-Zahra GE, Goetheer E (2019) Ship-based carbon capture onboard of diesel or LNG-fuelled ships. Int J Greenhouse Gas Control 85:1–10. https://doi.org/10.1016/j.ijggc.2019.03.008
Font-Palma C (2021) Is carbon capture a viable solution to decarbonise the shipping industry? In: Trondheim conference on CO_2 capture, transport and storage (TCCS-11), June 21–23 2021, Trondheim-Norway
Fradette L, Lefebvre S, Carley J (2017) Demonstration results of enzyme-accelerated CO_2 Capture. Energy Procedia 114:1100–1109. https://doi.org/10.1016/j.egypro.2017.03.1263
Gray ML, Soong Y, Champagne KJ, Pennline H, Baltrus JP, Stevens RW Jr, Khatri R, Chuang SSC, Filburn T (2005) Improved immobilized carbon dioxide capture sorbents. Fuel Process Technol 86:1449–1455
Güler E, Ergin S (2021) An investigation on the solvent based carbon capture and storage system by process modeling and comparison with another carbon control methods for different ships. Int J Greenhouse Gas Control 110:103438. https://doi.org/10.1016/j.ijggc.2021.103438
Heinz W (1982) The TOWS matrix—a tool for situational analysis. Long Range Plan 15(2):54–66
IMO (2019) Data collection system for fuel oil consumption of ships https://www.imo.org/en/OurWork/Environment/Pages/Data-Collection-System.aspx
IMO (2022). Energy Efficiency Measures https://www.imo.org/en/OurWork/Environment/Pages/Technical-and-Operational-Measures.aspx#:~:text=The%20EEDI%20provides%20a%20specific,parameters%20for%20a%20given%20ship.
International Chamber of Shipping (ICS) (2021) International Chamber of Shipping sets out plans for global carbon levy to expedite industry decarbonisation. https://www.ics-shipping.org/press-release/international-chamber-of-shipping-sets-out-plans-for-global-carbon-levy/
International Energy Agency (IEA) (2009) Technology roadmap, carbon capture and storage
International Organization for Standardization (ISO) 8217 (2017) Petroleym products - fuels (class F) - specifications of marine fuels
IPCC (2005) IPCC special report on carbon dioxide capture and storage. In: Metz B, et al. (ed) Prepared by working group III of the intergovernmental panel on climate change. Cambridge, United Kingdom and New York, NY, USA

Jajesniak P, Ali HEMO, Wong TS (2014) Carbon dioxide capture and utilization using biological systems: opportunities and challenges. J Bioprocess Biotech 3:1–15. https://doi.org/10.4172/2155-9821.1000155

Ji G, Zhao M (2016) Membrane separation technology in carbon capture, recent advances in carbon capture and storage. Yongseun Yun, IntechOpen. https://doi.org/10.5772/65723

Ji C, Yuan S, Huffman M, El-Halwagi MM, Wang Q (2021) Post-combustion carbon capture for a tank to propeller via process modeling and simulation. J CO_2 Util 51:101655. https://doi.org/10.1016/j.jcou.2021.101655

Jilvero H, Mathisen A, Eldrup N, Normann F, Johnsson F, Müller G, Melaaen MC (2014) Techno-economic analysis of carbon capture at an aluminum production plant - comparison of post-combustion capture using MEA and ammonia. Energy Procedia 63:6590–6601. https://doi.org/10.1016/j.egypro.2014.11.695

Kwon S, Fan M, DaCosta HFM, Russell AG, Berchtold KA, Dubey MK (2011) CO_2 sorption. Coal Gasif Its Appl 10:293–339

Lam MK, Lee KT, Mohamed AR (2012) Current status and challenges on microalgae-based carbon capture. Int J Greenhouse Gas Control 10:456–469. https://doi.org/10.1016/j.ijggc.2012.07.010

Luo X, Wang M (2017) Study of solvent-based carbon capture for cargo ships through process modelling and simulation. Appl Energy 195:402–413. https://doi.org/10.1016/j.apenergy.2017.03.027

Malmgren E, Brynolf S, Fridell E, Grahn M, Andersson K (2021) The environmental performance of a fossil-free ship propulsion system with onboard carbon capture - a life cycle assessment of the HyMethShip concept. Sustain Energy Fuels 5:2753. https://doi.org/10.1039/d1se00105a

MAN (2014) Waste heat recover system (WHRS) for reduction of fuel consumption, emissions and EEDI. MAN Technical Report

Markewitz P, Kuckshinrichs W, Leitner W, Linssen J, Zapp P, Bongartz R, Schreiber A, Müller TE (2012) Worldwide innovations in the development of carbon capture technologies and the utilization of CO_2. Energy Environ Sci 5:7281–7305. https://doi.org/10.1039/c2ee03403d

Medimurec NG, Mavar KN, Majic M (2018) Carbon capture and storage (CCS): technology, projects and monitoring review. The Mining-Gelology-Petroleum Engineering Bulletin 33:1–15. https://doi.org/10.17794/rgn.2018.2.1

Mikulcic H, Skov IR, Dominkovic DF, Alwi SRW, Manan ZA, Tan R, Duic N, Mohamad SNG, Wang W (2019) Flexible carbon capture and utilization technologies in future energy systems and utilization pathways of captured CO_2. Renew Sustain Energy Rev 114:1–20. https://doi.org/10.1016/j.rser.2019.109338

Mukherjee A, Okolie JA, Abdelrasoul A, Niu C, Dalai AK (2019) Review of post-combustion carbon dioxide capture technologies using activated carbon. J Environ Sci 83:46–63. https://doi.org/10.1016/j.jes.2019.03.014

Oh J, Anantharman R, Zahid U, Lee P, Lim Y (2022) Process design of onboard membrane carbon capture and liquefaction systems for LNG-fueled ships. Sep Purif Technol 282 Part B:120052. https://doi.org/10.1016/j.seppur.2021.120052

Pocklington D, Leese R (2009) Towards zero emission production - potential of carbon capture in energy intensive industry. Environ Sci Technol 29:126–154. https://doi.org/10.1039/9781847559715-00126

Raza A, Rezaee R, Gholami R, Bing CH, Nagarajan R, Hamid MA (2016) A screening criterion for selection of suitable CO_2 storage sites. J Nat Gas Sci Eng 28(2016):317–327. https://doi.org/10.1016/j.jngse.2015.11.053

Raza A, Gholami R, Rezaee R, Rasouli V, Rabiei M (2019) Significant aspects of carbon capture and storage - a review. Petroleum 5:335–340. https://doi.org/10.1016/j.petlm.2018.12.007

Repasky J, McCarthy D, Armstrong P, Carolan (2014) ITM technology for carbon capture on natural gas and hybrid power systems

Salvi BL, Jindal S (2019) Recent developments and challenges ahead in carbon capture and sequestration technologies. SN Applied Sciences 1:885. https://doi.org/10.1007/s42452-019-0909-2

Sandhyarani N (2019) Surface modification methods for electrochemical biosensors. Electrochem Biosen 3:45–75. https://doi.org/10.1016/B978-0-12-816491-4.00003-6
Song C, Liu Q, Deng S, Li H, Kitamura Y (2019) Cryogenic-based CO_2 capture technologies: state-of-the-art developments and current challenges. Renew Sustain Energy Rev 101:265–278. https://doi.org/10.1016/j.rser.2018.11.018
Thambimuthu K, Davison J, Gupta M (2002) CO_2 capture and storage Proceedings of IPPC workshop on carbon dioxide capture and storage. 18–21 November 2002, Regina-Canada
The International Council on Clean Transportation (ICCT) (2018) The international maritime organization's initial greenhouse gas strategy
United Nations Conference on Trade and Development (UNCTAD) (2021) Review of maritime transport 2021
Ünveren EE, Monkul BÖ, Sarıoğlan Ş, Karademir N, Alper E (2017) Solid amine sorbents for CO_2 capture by chemical adsorption: a review. Petroleum 3(1):37–50. https://doi.org/10.1016/j.petlm.2016.11.001
Vega F, Cano M, Camino S, Fernandez LMG, Portillo E, Navarrete B (2017) Solvents for carbon dioxide capture. Carbon dioxide chemistry, capture and oil recovery. Iyad Karame, Janah Shaya and Hassan Srour, IntechOpen.https://doi.org/10.5772/intechopen.71443
Wang J, Ryan D, Anthony E, Wildgust N, Aiken T (2011) Effects of impurities on CO_2 transport, injection and storage. Energy Procedia 4:3071–3078. https://doi.org/10.1016/j.egypro.2011.02.219
Ye B, Jiang J, Zhou Y, Liu J, Wang K (2019) Technical and economic analysis of amine-based carbon capture and sequestration at coal-fired power plants. J Clean Prod 222:476–487. https://doi.org/10.1016/j.jclepro.2019.03.050
Zaman M, Lee JH (2013) Carbon capture from stationary power generation sources: a review of the current status of the technologies. Korean J Chem Eng 30:1497–1526. https://doi.org/10.1007/s11814-013-0127-3
Zhou P, Wang H (2014) Carbon capture and storage-Solidification and storage of carbon dioxide captured on ships. Ocean Eng 91:172–180. https://doi.org/10.1016/j.oceaneng.2014.09.006
Zincir B (2019) An alternative fuel assessment model for ships and experiments on the effect of methanol on diesel engines. Dissertation thesis. Department of Maritime Transportation Engineering, Maritime Faculty, Istanbul Technical University
Zincir BA (2020) Comparison of the carbon capture systems for onboard application and voyage performance investigation by a case study. M.Sc. Thesis. Department of Maritime Transportation Engineering, Maritime Faculty, Istanbul Technical University

Part IV
Green Port Concept

Chapter 8
Green Concept of Ports and Transition Model

M. Ziya Sogut

Abstract Ports with logistics mobility have a critical role in the green transition and decarbonization targets of the sector. The decarbonization determination of maritime transport, supported by the COP 26 sectoral call and the 2050 targets, was expressed in a structure that also includes ports. In this context, ports need new strategic approaches to support port authorities in terms of energy and environmental sustainability. For this purpose, a new approach, which is defined as the "green concept framework," was developed in this study, together with the sustainability principles of ports. The study was handled from two perspectives in terms of energy management and sustainability principles based on the green concept over the reference port area. A holistic energy efficiency potential for reference ports and 13 criteria for the green concept evaluation were examined. The analyses identified primarily energy use and waste management as the main problem areas for the port. In addition, the areas that need to be improved were defined by the port authorities based on ISO 50001 principles. In the analysis made, suggestions were developed to support energy and environmental sustainability in ports depending on the emission savings potential of 8,21%.

Keywords Ports · Green concept · Energy · Environmental · Sustainability

Nomenclature

AHP	Analytic hierarchy process
COP26	UN Climate Change Conference of the Parties
CR	Consistency ratio
ESPO	European Maritime Port Organization
HDD	Heating degree day
IMO	International Maritime Organization

M. Z. Sogut (✉)
Piri Reis University, Maritime Faculty, Istanbul, Türkiye
e-mail: mzsogut@gmail.com; mzsogut@pirireis.edu.tr

B. Zincir et al. (eds.), *Decarbonization of Maritime Transport*, Energy, Environment, and Sustainability, https://doi.org/10.1007/978-981-99-1677-1_8

8.1 Introduction

In social sustainability, the protection of environmental structures and especially the reduction of fossil fuel-induced global warming is possible with the joint contributions of all sectors. In this context, this process, which is handled by many disciplines in the maritime sector as well as in the ports, many conceptual orientations have been developed, primarily sustainability. As a matter of fact, the maritime sector, with the regulations developed by the International Maritime Organization (IMO), has brought forward the concepts of resource management and sectoral sustainability in all aspects (Woo et al. 2018). In terms of sectoral targets, it is valuable in terms of development strategies of environmental sustainability for the maritime sector, which is responsible for approximately 85% of the global trade volume in the transportation sector (UNCTAD 2018). In this context, sectoral frameworks have been developed and green concept such as decarbonization is developing as an important step for the maritime sector. The maritime industry is a complex structure consisting of many operational processes. In this process, especially for this process where pollution is a great potential structure besides resource management, ports are components that need to be managed multi-dimensionally with their operational processes (Tichavska et al. 2017). In fact, ports are structures that should be based on sustainability as organizations living in corporate processes.

Sustainable port concept; As a strategic idea, it can be defined as a development that brings new balances in terms of environment and development, with its ideology, innovative technology, business strategies based on efficiency, and aiming at continuous development, including stakeholders. This conceptual development can be seen as a paradigm shift that primarily prioritizes sustainability for port authorities and authorities. The global economies expect that depending on the trade growth scenarios, the investments in ship sizes and related port facilities and capacities will increase (OECD 2012; PIANC 2014). In addition, depending on the increase in sectoral traffic, it is predicted that the sectoral economic potential will continue to grow rapidly (Zdravev 2017; Lam ve Notteboom 2012). While ports play a role in this growing economy, it has been observed that the demographic structure of port cities parallels this and affects the local hinterland (Ducruet and Lee 2006). It has been emphasized that in sustainable economies, it is necessary to develop regional and national economies and social structures in quantitative and qualitative terms with more sustainable models (Asgari et al. 2015). In this context, ports have developed as a paradigm based on change, not only for maritime transport but also for the settlement areas where they are located. It refers to the development of many operations and related actions based on this change, especially for green ports that emphasize sustainability (WG150 2013). In particular, the concept of green, which has gained global value, has developed new trade areas that form the infrastructure for circular economies, together with its environmental priorities (Mintu and Lozada 1993). As a matter of fact, this paradigm shift, which port authorities attach importance to in the maritime sector, has been a key that port administrations care about to improve sustainability. However, studies showing that conceptual change is a competitive feature for port economies are also noteworthy (Lai et al. 2013).

The green port should be considered fundamentally as a conceptual structure. Green port, which has been included in the literature with different definitions for the last 20 years, actually dates back to the 1990s. The first institutional examples were developed in Europe, and the first study of the European Maritime Port Organization (ESPO), known as the Environmental Code of Practice, was published in 1994. Later, at the end of two revision studies, such as in 2003 and 2012, the basic parameters of the transformation for ports known as the ESPO Green Guide were updated and published. Green port applications, which have gained a standard profile, have found application areas for many ports in the world, such as Shanghai, Hong Kong, and Singapore Ports. Today, the concept of the green port has been discussed in many ways. These are emissions and port management (Dragovic et al. 2015), environmental management and policies (Galeotti et al. 2018), green ports (Barnes-Dabban et al. 2017), and port emissions and reduction (Ratanavaraha and Jomnonkwao 2015; Yang 2017).

The green port concept is a structural process directly related to many parametric functions, especially port operations. In this context, first of all, it is necessary to develop holistic evaluation criteria by port authorities. In this context, three basic questions were taken into consideration for this process, which is seen as a deficiency in the literature.

1. Can a holistic framework be defined for green port applications?
2. Can a structural analysis model be developed based on sustainability in green ports?
3. Can the priority criteria or criteria be defined for green port applications that should be taken into account by the authorities or administrations?

This study presents an approach framework developed for the reference port group. In particular, the energy efficiency of the ports and their related effects were discussed primarily through the energy and environmental sustainability criteria taken as the basis for port applications. Then, an impact assessment based on the AHP analysis was carried out on the defined green port criteria.

8.2 Port Management

In global trade, ports, especially international transportation, have become valuable with their preferences to manage regional advantages. The increase in maritime transportation and the pressure of transportation cost effects on competition necessitated the development of port infrastructures and the provision of technological advantages. Changes in the management and service structures of the ports, especially the ports on the trade corridors, have been inevitable. Environmental sustainability is a valuable parameter for circular economies. In this economic framework, having large port structures and meeting sufficient cargo volume in terms of global maritime is a growth-oriented expectation. This directly results in fewer ship cycles for ports along with ease of handling while reducing transit inventory costs. All these processes are a critical choice for economies of scale in global competition (Esther 2007). For a key of such high importance, the development of management organizations is an institutional need. This structural need, together with environmental factors, necessitates the establishment of a framework by the port authorities while guiding national policies.

Ports should have a dynamic management structure together with operational management elements. In particular, it consists of port and terminal structures and strategic decision structures based on cargo handling, storage, facilities, and processes. In this respect, managerial strategies in ports include objectives developed for processes and operations (Magdiel et al. 2022). Administrative processes in ports should develop focal solutions based on optimal solutions. As a matter of fact, as in the literature, Bierwirth and Meisel (2010) examined studies based on different problems in quay operations and developed solutions. Bjerkan and Seter (2019) examined port infrastructures within the scope of sustainability. According to Campisi et al. (2022), who reviewed over 70 studies between 2010 and 2018, port management focus on sustainability; port management, energy, maritime, and land activities.

Along with the global changes, port management has gained value in considering a change that takes into account the green concept as an economic and environmental requirement in order to improve sustainability. In this context, it can be counted to reduce environmental threats in operational processes, along with energy use, especially in maritime and land activities (Cheng et al. 2013; Lirn et al. 2013). The manageability of port processes today is based on a change within management technologies. In this context, port management should also develop this change. As a matter of fact, the use of smart technology tools with global networks has developed as a necessity. In this respect, port administrations are expected to develop steps that can manage change by developing focal points, as shown in Fig. 8.1.

To improve attention to the environmental impacts of maritime exercises, as well as to reduce the ecological damage created by transportation operations, such as atmospheric changes (Cheng et al. 2013; Lirn et al. 2013). Accordingly, ports and their stakeholders can use their green initiatives to seek business profits. Expanding cargo volume in a port can bring with it the problem of environmental and ecological protection.

8.3 Green Concept and Sustainability

In today's world, the reduction of energy and natural resources is not only a regional problem but a global problem that affects the whole world. In this respect, the development of holistic models comes to the fore in development goals for all elements required by social development, especially resource management. Ports are not structures to be critiqued only in terms of the cities they are located in. It is an important

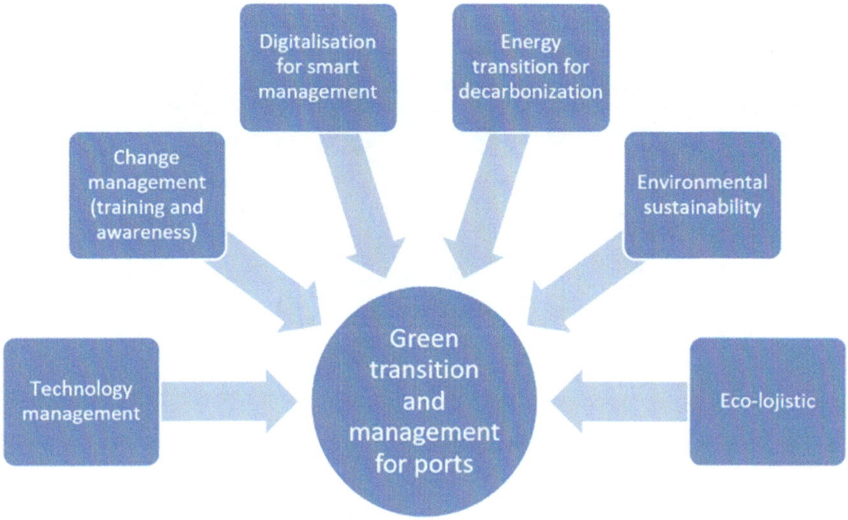

Fig. 8.1 Main pillars for the green port concept

key point, especially in maritime transport, and is the cornerstone of development for the cities where they are located. As living organizations, ports are an organic part of cities and their development should be handled with a sustainability approach.

From a sustainability perspective, a port must manage and balance the three bottoms. economic welfare, social welfare, and environmental quality (Lam and Yap 2019). An important driver for sustainability is attracting and retaining customers who value sustainability (Chan et al. 2012). As such, a tangible commitment to proactive corporate orientation and environmental performance can be a business opportunity for growth. In fact, today, the sustainability of ports should be considered together with three components: technology, efficiency, and the environment. These processes should be integrated into a holistic model together with the social, economic, and environmental factors in the development process of the ports. However, today, the ports and port areas that have grown over time have turned into problem areas with uncontrolled energy consumption, taking up more space, and destroying nature. Social pollution, including traffic on logistics lines, can also be involved.

Today, there is no standard framework developed in the direction of green shipyards. However, important frameworks on this subject have been developed in the literature. For example, Dangelico and Pontrandolfo (2010) developed this according to the impact approach in the life cycle process. In this context, the relationship between the business processes and the environment of an environmentally friendly shipyard should be defined at every stage, and this focus should be determined for the shipyard. Studies have identified materials, energy, and pollution as key indicators, especially for the environmental impact point (EED 2012; Rijksdienst 2016). However, considering the operational processes in the shipyards, the structural potentials of the main indicators in terms of impact potentials should be categorized

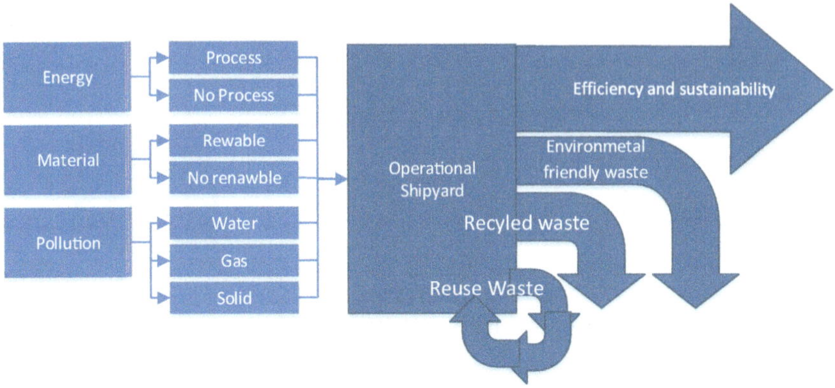

Fig. 8.2 Green flow framework of a shipyard

according to their functions. Especially for the zero impact point, energy and pollution are indicators that should be studied. The green flow framework developed in this context defines a direction of movement for shipyards. As can be seen in Fig. 8.2, this includes process evaluations on energy, materials, and pollution.

8.4 Methodology

The green port concept requires an approach that must be considered within a framework. In this respect, the green concept shows a framework that also expresses a search since the 2000s. The green concept is basically handled within a behavioral or institutional structure. It describes the effects of this, the management of costs, and the fundamental relationships of green transformation. In the first part of the study, the need for a holistic framework for the green port is defined. In this context, the methodological framework developed is planned as an applicable concept in ports. The indicators needed in the processes and implementation processes related to the use of energy, which is defined in flow processes, are discussed in the result and discussion section of the study. In Fig. 8.3, a holistic framework was developed for the relational connections of the parameters defined in this scope.

The green concept should be considered as a process of transformation. On this basis, it is necessary to see a change that affects behavior and institutional demands. The manageability of energy defines a structural control structure for ecological, economic, and sustainability. In this respect, it is necessary to see a structure that adds value to corporate strategies. In this respect, the green concept should be considered as a managerial process for corporate structures. Basically, it presents a model that enables awareness to turn into behavior together with resource, environment, and technology management.

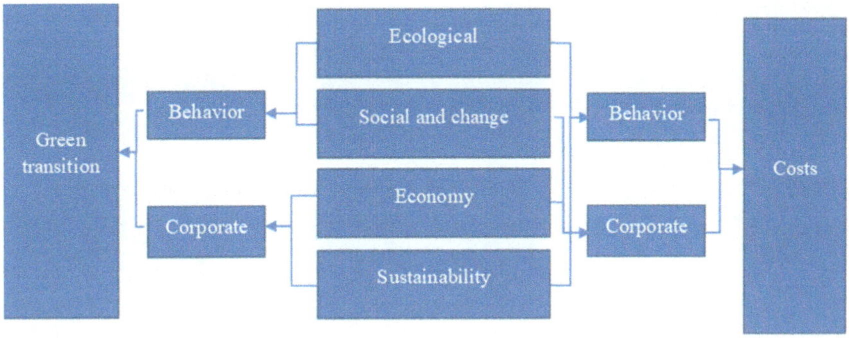

Fig. 8.3 Framework of green concept

The use of resources should gain a sustainable value, especially energy. For this purpose, energy management is a priority concept in corporate structures and requires a holistic approach model in energy management, especially in port operations. On the basis of the green concept, energy and resource management are a priority step. In particular, defining inefficiency, reducing fossil fuel consumption or creating alternatives, and choosing renewable energy sources can be seen as a need. While green concept work gains importance for corporate strategies, it is a process where priorities should be evaluated. Today, such definitions can be seen. As a matter of fact, ESPO defines priorities for environmental strategies in this context every year. Figure 8.4 has defined 2019 priorities for ports in this direction.

All these approaches should emphasize a strategic framework for faith management in operational processes. For this purpose, a structural analysis model has been developed to define the concept of green in ports. The flow processes of this model are used as a flow methodology for green ports in the study, and the frame of this model is given in Fig. 8.5.

The green concept framework developed in this study is handled in a two-stage structure and evaluated depending on a reference port in Istanbul/Turkey. The port is located in a sheltered area in terms of physical conditions and has a logistics flow connected with roads and railways. While the annual average sea water temperature for Istanbul varies between 16–17 °C and the yearly outdoor temperature varies between 6/28 °C, the region can reach 50 knots under the influence of southwest winds. In the region with 6–7 miles of sea current, it has a medium-sized structure with a port span of 2765 m and a pier capacity of 8.5 million tons/year. The port, which has a wide range of structural services including ship, cargo, and related services, has the authority of a licensed waste reception facility. The port, which has a total of 89 cranes and handling equipment, has a total consumption of 4.1 GWh/year, including electricity, natural gas, and diesel fuel.

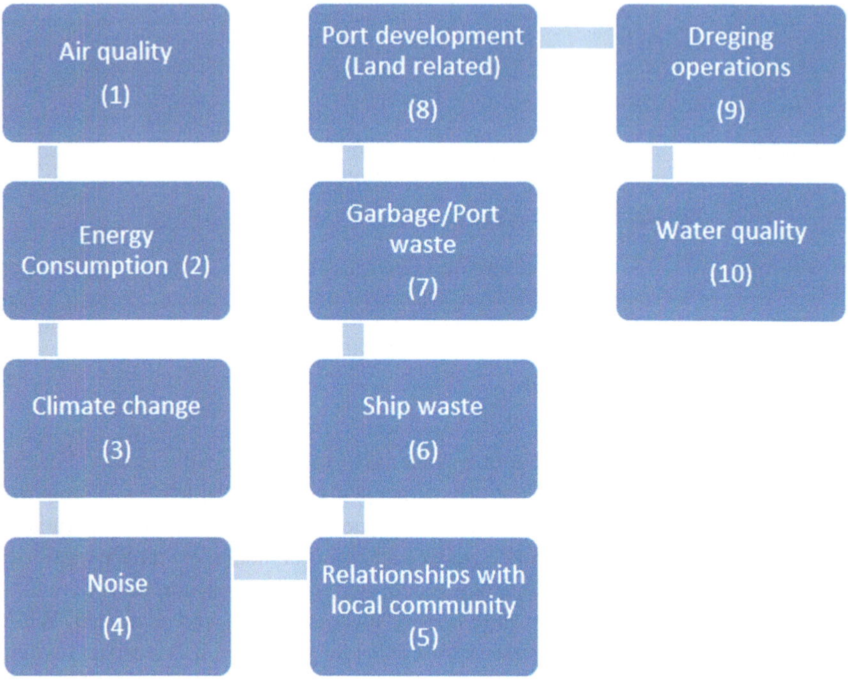

Fig. 8.4 Environmental priorities of Europe ports (2019)

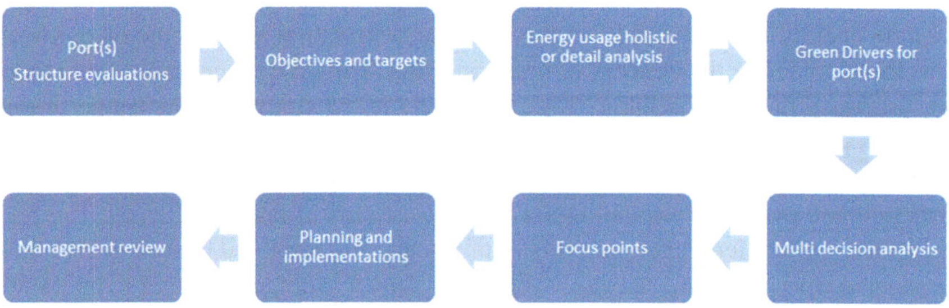

Fig. 8.5 Methodologic framework for Green concept

8.5 Results and Discussion

In this study, the analyses are based on the evaluation of a port with indicators based on the situation analysis of energy use for the decision processes of the Green Concept. First of all, the status assessment of the port according to ISO 50001 was discussed, and then the efficiency potential to improve the action behaviors of energy management was examined. The GAP Analysis is an intuitive approach created by the opinion of direct interest in the subject or process in which it is applied. In this study, port-related expert opinions were evaluated by taking. The GAP analysis of the port was carried out according to the ISO50001 energy management system in Fig. 8.6, and the management infrastructure was evaluated.

While the ISO9001 quality and 14,001 environmental management systems are being implemented in the port, there is a lack of energy management. The establishment of managerial stability and related management infrastructure for an institutional structure is a primary need. The sustainable energy efficiency of the port has been discussed over the last three years' data and its efficiency performance has been evaluated. In this context, regression analysis was performed and the results are given in Fig. 8.7.

Fig. 8.6 GAP analysis of port based on ISO 50001

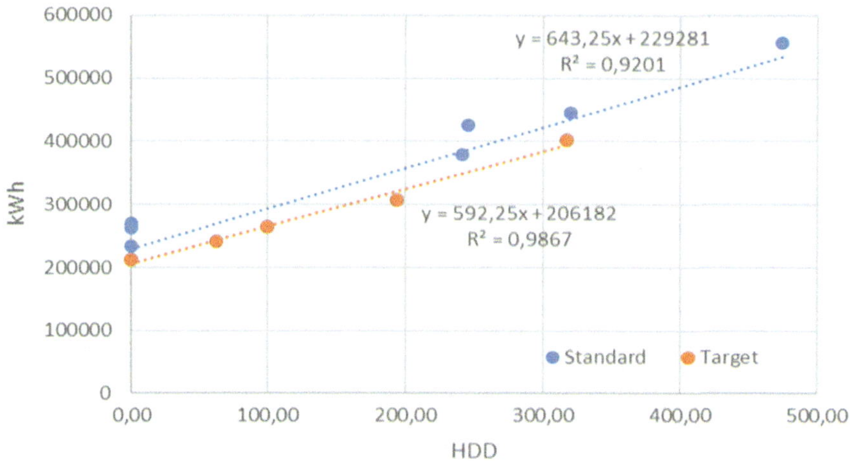

Fig. 8.7 Energy efficiency potential of port

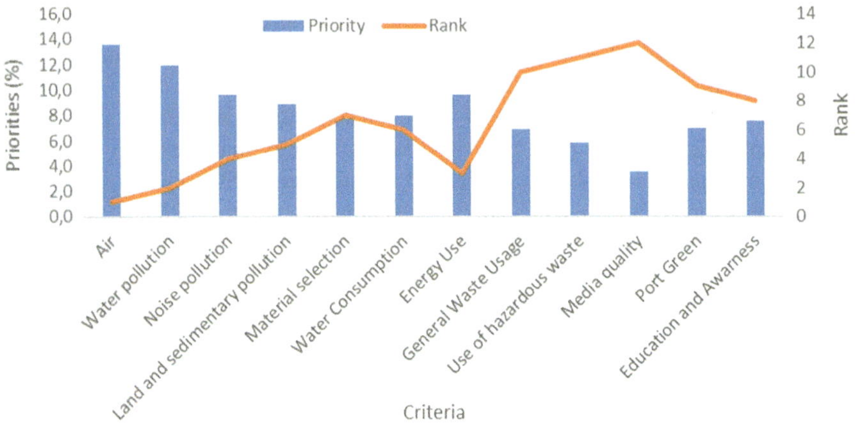

Fig. 8.8 Priorities distribution of criteria

A regression analysis has been conducted over the annual consumption of the port due to electricity, natural gas, and diesel consumption, and it has been examined over the last three years. It was seen that the best variable, depending on the consumption distribution for the first year, was the HDD value, and the target was defined by considering the standard three energy potentials. According to the regression analysis, the R2 value for the last two years was 98.67%, and the total efficiency potential was found to be 8.21%. The third phase of his work is to determine the priorities for the green transformation of the port, depending on the green concept. At this stage, an evaluation was made on 12 criteria, which means the road map for the structural change of the port. The interrelationship matrix for reference values is graded from 1 to 9 in the range of equal importance and extreme importance. The distributions based on the analyses made in this context are given in Fig. 8.8.

According to the analysis made, air management in ports was seen as an effective criterion, while water pollution and energy use were found to be 11.9% and 9.6%, respectively. In the analysis, the decision matrix was examined, and the developed matrix table was presented as Table 8.1.

8.6 Conclusion

This study is based on a model framework developed for ports based on the green concept. In this framework, priorities based on the AHP analysis for energy management assessment and green transition for ports were evaluated. In this context,

a. The GAP Analysis showed the Port's need for Effective Energy Management. In particular, the development of energy users and the development of energy action plans can be defined as institutional needs.

Table 8.1 Decision matrix of criteria

	1	2	3	4	5	6	7	8	9	10	11	12
1	1	2.00	5.00	1.00	2.00	2.00	1.00	2.00	1.00	3.00	1.00	1.00
2	0.50	1	3.00	2.00	3.00	1.00	1.00	2.00	2.00	3.00	1.00	1.00
3	0.20	0.33	1	2.00	2.00	3.00	1.00	1.00	2.00	3.00	1.00	1.00
4	1.00	0.50	0.50	1	1.00	2.00	1.00	3.00	1.00	3.00	1.00	1.00
5	0.50	0.33	0.50	1.00	1	2.00	1.00	1.00	2.00	3.00	1.00	1.00
6	0.50	1.00	0.33	0.50	0.50	1	1.00	3.00	2.00	3.00	1.00	1.00
7	1.00	1.00	1.00	1.00	1.00	1.00	1	2.00	2.00	3.00	2.00	1.00
8	0.50	0.50	1.00	0.33	1.00	0.33	0.50	1	2.00	3.00	2.00	1.00
9	1.00	0.50	0.50	1.00	0.50	0.50	0.50	0.50	1	2.00	1.00	1.00
10	0.33	0.33	0.33	0.33	0.33	0.33	0.33	1.33	1.50	1	1.00	1.00
11	1.00	1.00	1.00	1.00	1.00	1.00	0.50	0.50	1.00	1.00	1	1.00
12	1.00	1.00	1.00	1.00	1.00	1.00	1.00	1.00	1.00	1.00	1.00	1

The decision matrix based on priorities was found as Number of comparisons = 66, Consistency Ratio CR = 7.1%, Principal eigen value = 13,197 and Eigenvector solution: 6 iterations, delta = 7.5E-9

b. Based on the regression analysis, the energy effective potential of the port was found to be 8.21%. This value is the target value developed for energy management processes and will provide the definition of the actions that energy management should develop for energy users.
c. Air management, water pollution, and energy use are the three main criteria for a green port transition. These criteria are important in terms of defining corporate goals for the defined port.

It is important to create an institutional structure in energy and environmentally sustainable ports. Energy management in a structure is also a requirement. In this context, the creation of the sustainability unit of these processes in ports and the manageability of the processes in this context will be appropriate. However, process impact analyses for the establishment of the Energy Management Infrastructure and action strategies are first recommended. These studies should be developed for optimization for specific energy users in energy consumption and studies on boundary conditions for defined indicators.

References

Asgari N, Jones D, Hassani A, Nguyen H (2015) Sustainability ranking of the UK major ports: methodology and case study. Transp Res E-Log 78:19–39

Barnes-Dabban H, vanTatenhove JPM, vanKoppen KCSA, Termeer KJAM (2017) Institutionalizing environmental reform with sense-making: West and Central Africa ports and the 'green port' phenomenon. Mar Policy 86:111–120

Bjerkan KY, Seter H (2019) Reviewing tools and technologies for sustainable ports: does research enable decision making in ports? Transport Res Transport Environ 17:243–260

Campisi T, Samuele M, Giacomo C, Luca L, Sabrina M, Francesco M, Davide S (2022) Locally integrated partnership as a tool to implement a smart port management strategy: the case of the port of Ravenna (Italy), Ocean & Coastal Management, Volume 224. ISSN 106179:0964–5691. https://doi.org/10.1016/j.ocecoaman.2022.106179

Christian B, Frank M (2010) A survey of berth allocation and quay crane scheduling problems in container terminals. Europ J Oper Res 202(3):615–627

Chan RYK, He H, Chan HK, Wang WYC (2012) Environmental orientation and corporate performance: the mediation mechanism of green supply chain management and moderating effect of competitive intensity. Ind Mark Manag 41(4):621–630

Chen G, Govindan K, Golias MM (2013) Reducing truck emissions at container terminals in a low carbon economy: proposal of a queueing-based bi-objective model for optimizing truck arrival pattern. Transp Res Part e: Logist Transp Rev 55:3–22

Dangelico RM, Pontrandolfo P (2010) From green product definitions and classifications to the Green Option Matrix. J Clean Prod 18(16):1608–1628

Dragovic B, Tzannatos E, Tselentıs V, Mestrovic R, Skuric M (2015) Ship emissions and their externalities in cruise ports. Transp Res Part D Transp Environ 61:289–300

Ducruet C, Lee S-W (2006) Frontline soldiers of globalisation: port-city evolution and regional competition. GeoJournal 67(2):107–122

EED (2012) Directive 2012/27/eu of the european parliament and of the council of 25, Energy Efficiency Directive October 2012 on energy efficiency, amending directives 2009/125/ec and 2010/30/eu and repealing directives 2004/8/ec and 2006/32. Official J L 315:1–56

Esther RS (2007) Ports hierarchy: structure and new demands. Enfasis Logistica VII:40–51

Galeotti M, Rubashkina Y, Salini S, Verdolini E (2018) Environmental policy performance and its determinants: application of a three-level random intercept model. Energy Policy 114:134–144

Lai KH, Lun YHV, Wong CWY, Cheng TCE (2013) Measures for evaluating green shipping practices implementation. Int J Shipp Transp Logist (IJSTL) 5(2):217–235

Lam JSL, Notteboom T (2012) The green port toolbox: a comparison of port management tools used by leading ports in Asia and Europe. Paper presented at ınternational association of maritime economists (IAME) conference, Taipei, Taiwan, 5–8 September 2012

Lam JSL, Yap WY (2019) A stakeholder perspective of port city sustainable development. Sustainability 11:447

Lirn T, Wu YJ, Chen YJ (2013) Green performance criteria for sustainable ports in Asia. Int J Phys Distrib Logist Manag 43(5/6): 427–45

Merkel, A., Kalantari, J., Mubder, A. (2022). Port call optimization and CO2-emissions savings – Estimating feasible potential in tramp shipping. Maritime Transport Research Magdiel A. Agüero-Tobar, Marcela C. González-Araya, Rosa G. González-Ramírez, 2022Merkel A, Kalantari J, Mubder A (2022) Port call optimization and CO_2-emissions savings—estimating feasible potential in tramp shipping. Maritime Transport Research Magdiel, Agüero-Tobar A, González-Araya MC, González-Ramírez RG (2022) Assessment of maritime operations efficiency and its economic impact based on data envelopment analysis: a case study of Chilean ports. Res Transp Bus Manag, 100821, ISSN: 2210-5395. https://doi.org/10.1016/j.rtbm.2022.100821

Mintu AT, Lozada HR (1993) Green marketing education: a call for action. Mark Educ Rev 3 (Fall), 17–23 PIANC, 2014b. 'Sustainable Ports' A Guide for Port Authorities. PIANC report no 150–2014

OECD (2012). http://www.oecd.org/regreform/sectors/48837794.pdf.

Ratanavaraha V, Jomnonkwao S (2015) Trends in Thailand CO_2 emissions in the transportation sector and Policy Mitigation. Transp Policy 41:136–146

Rijksdienst voor O (2016) The energiemanagementsysteem. Nederland. [Online]. Available: http://www.rvo.nl/subsidies-regelingen/meerjarenafspraken-energie-effici%C3%ABntie/verplichtingen-ja3/mee/energiemanagement/softwaresystemen/het-energiemanagementsysteem

Tichavska M, Tovar B (2017) External costs from vessel emissions at port: a review of the methodological and empirical state of the art. Transp Rev 37(3):383–402, ISSN 0144-1647. https://doi.org/10.1080/01441647.2017.1279694

UNCTAD (2018) Review of maritime transport 2018. United Nations Conference on Trade and Development, Geneva

WG150 (2013) WG150 'sustainable ports' a guidance for port authorities. The world association for waterborne transport ınfrastructure revision 6

Woo JK, Moon DSH, Lam JSL (2018) The impact of environmental policy on ports and the associated economic opportunities. Transport Res Part A 110:234–242

Yang YC (2017) Operating strategies of CO_2 reduction for a container terminal based on carbon footprint perspective. J Clean Prod 141:472–480

Zdravev AZ (2017) Port sustainability practices: a case study of port of rotterdam and port of los angeles. Port and Transport Economics Erasmus University Rotterdam, Department of Urban

Chapter 9
Evaluation of the Green Port Concept for Decarbonized Maritime Industry

İsmail Kurt

Abstract Toward decarbonization of entire maritime transport chain, it is important to develop not only greener ships but also port and maritime logistics systems that work with alternative energy sources providing significant environmental and economic benefits. Ports are essential energy consumers and one of the key elements of the maritime transport chain. The green port concept, on the other hand, is an approach aiming to minimize fossil fuel consumption to reach more environmentally friendly and economically sustainable port operations. The aim of this chapter is to investigate different perspectives on how to make sustainability assessment in ports. The chapter defines a green port development model by evaluating new green technologies, low or zero-carbon alternative terminal equipment, and other energy-consuming components of ports. To achieve these aims, environmentally friendly green port applications in all forms are being reviewed on a global scale. In green port applications, issues such as energy conservation, air pollution, water pollution, hazardous substances management, and habitat are specially pointed out due to their direct relevancy with port operations. This chapter reveals the contribution of green port practices to energy efficiency, environmental protection, and sustainable industrial and regional development. The chapter also draws attention to the importance of environmental awareness in ports, especially located in or nearby urban areas.

Keywords Green port operations · Sustainable shipping · Alternative energy

Abbreviations

APSN	APEC Port Services Network
BUNKER	International Convention on Civil Liability for Bunker Oil Pollution Damage
BWM	International Convention for the Control and Management of Ships' Ballast Water and Sediments

İ. Kurt (✉)
Faculty of Naval Architecture and Maritime, Izmir Katip Celebi University, Izmir, Turkey
e-mail: ismail.kurt@ikc.edu.tr

B. Zincir et al. (eds.), *Decarbonization of Maritime Transport*, Energy, Environment, and Sustainability, https://doi.org/10.1007/978-981-99-1677-1_9

CLC	International Convention on Civil Liability for Oil Pollution Damage
ECA	Emission Control Areas
EMS	Environmental Management System
ESPO	European Sea Ports Organization
EU	European Union
GHG	Greenhouse gases
GPA	Global Program of Action for the Protection of the Marine Environment from Land-based Activities
GPAS	Green Port Award System
HFO	Heavy fuel oil
IAPH	International Association of Ports and Harbors
IMO	International Maritime Organization
IPCC	Intergovernmental Panel on Climate Change
ISO	International Organization for Standardization
LED	Light Emitting Diode
LNG	Liquefied natural gas
LPG	Liquefied petroleum gas
MARPOL 73/78	International Convention for the Prevention of Pollution from Ships
MGO	Marine gas oil
MPEC	Marine Environmental Protection Committee
NOX	Nitrogen oxides
OHSAS	Occupational Health and Safety Assessment Series
OPRC	International Convention on Oil Pollution Preparedness, Response and Co-operation
OPS	Onshore power supply
PM	Particulate material
QMS	Quality Management System
SECA	Sulfur Emission Control Areas
SOLAS	International Convention for the Safety of Life at Sea
SOX	Sulfur Oxide
SSE	Shore-side electricity
UN	United Nations
UNCLOS	United Nations Convention on the Law of the Sea
UNCTAD	United Nations Conference on Trade and Development
UNEP	United Nations Environment Program
WMO	World Meteorological Organization
WPCI	World Port Climate Initiative
WPSP	World Port Sustainability Program

9.1 Introduction

Ports are one of the key components of the global supply chains serving the basic principles of globalization which are predominantly low-cost production, transportation, and outsourcing. According to UNCTAD (2021) report (2021), ports are central nodes of global supply chains, with their role in transporting more than 80% of global trade by volume. As an important component of international trade, ports also contribute significantly to the development of many industries and cities around them.

Although the significant contributions of ports to economic growth and development, environmental concerns of port activities are increasing, especially by considering the location of ports for coastal zones, low-lying areas, and deltas. The environmental impacts of the ports are more important in terms of climate change due to the population density in the areas where the ports come into contact, the volume of trade activities carried out at the ports, and the size of the port structures. Associating the environmental impacts from port operations with the port hinterland may lead to more serious environmental consequences.

Nowadays, many green port themed projects and programs are carried out by governments and international organizations to ensure the environmental sustainability of ports (The European Sea Ports Organisation n.d.; VERİFAVİA n.d.; Ministry of Transport and Infrastructure of the Republic of Turkiye Green Port n.d.). Ports that successfully perform environmental management are also rewarded. At a period when the influences of climate change are clearly felt, incentives to reduce or eliminate port-related environmental impacts are as seen important in terms of increasing the environmental awareness of the port industry. In addition, it is possible to say that the success of the port authority in environmental management will return to commercial success, as ports with the title and logo of green port or eco-port can be preferred more by institutions and individuals.

Environmental concerns about ports generally focus on ship and cargo operations, logistics and industrial activities, and plans for port construction, development, and expansion (Luo and Yip 2013; Puig et al. 2015; Lam and Notteboom 2014). Firstly, environmental concerns from ship and cargo operations are related to air, water, and noise pollution. Secondly, while environmental concerns arising from logistics and industrial activities are linked with port-centered air and noise pollution, there are also environmental concerns due to the traffic density that may be experienced in railway and road connections. Finally, the environmental aspects that need to be addressed in port construction, development, and expansion are much more comprehensive. These are generally water quality, coastal hydrology, soil pollution, marine and coastal ecology, air quality, noise and vibration, waste management, visual deformation, and socio-cultural aspects. On the other hand, European Sea Ports Organization (ESPO) listed the top 10 environmental issues prioritized by European port authorities for the last 5 years as shown in Table 9.1 (The European Sea Ports Organisation n.d.).

Debate continues about defining and identifying the best port governance model for green port management (Munim et al. 2020). Although the environmental issues that ports prioritize can be identified, a green port management should meet not only

Table 9.1 Top 10 environmental priorities of EU ports, 2017–2021. (*Source* ESPO (n.d.))

	2017	2018	2019	2020	2021
1	Air quality	Air quality	Air quality	Air quality	Air quality
2	Energy efficiency	Energy efficiency	Energy efficiency	Climate change	Climate change
3	Noise	Noise	Climate change	Energy efficiency	Energy efficiency
4	Water quality	Relationship with local community	Noise	Noise	Noise
5	Dredging operations	Ship waste	Relationship with local community	Relationship with local community	Relationship with local community
6	Garbage/port waste	Port development (land related)	Ship waste	Ship waste	Water quality
7	Port development (land related)	Climate change	Garbage/port waste	Water quality	Ship waste
8	Relationship with local community	Water quality	Port development (land related)	Garbage/port waste	Dredging operations
9	Ship waste	Dredging operations	Dredging operations	Dredging operations	Port development (land related)
10	Climate change	Garbage/port waste	Water quality	Port development (land related)	Garbage/port waste

environmental objectives but also economic objectives for sustainable development. In line with these objectives, the role of the port, corporate social responsibility, national and international environmental regulations, and the characteristics of port users should be considered in determining the best governance strategy for green port management (Du et al. 2019). Therefore, since the dynamics are different for each port, it is not possible to talk about a "one size fits all" port governance model, and practices for green port management need to be analyzed on a case-by-case basis. To focus on the green port concept, the components of port operations that work like the gears of a mechanism should be analyzed in detail. The activities of these components can be classified in 3 categories as the causes of green port environmental concerns, as in Fig. 9.1 (Axel 2011). In this context, this chapter examines the green port practices used to demonstrate the green port approach in detail.

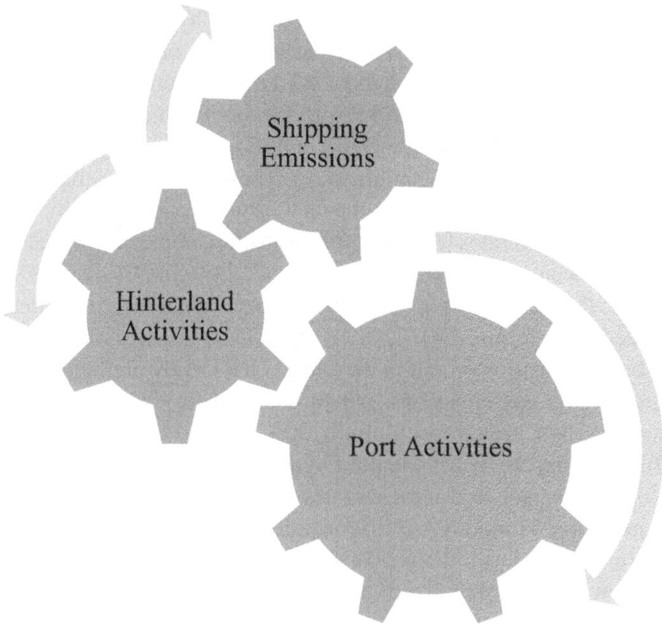

Fig. 9.1 Categories of green port environmental concerns

9.2 Green Port Concept

The concept of green port (also known as ecological port or eco-port) can be defined as encouraging and integrating all port workers and stakeholders into the eco-friendly development and operations of the port facility to increase sustainable environmental awareness. While providing environmental sustainability in green ports, sustainable ports emerge when a balance between social impacts and economic interests is considered (Oniszczuk-Jastrząbek et al. 2018; Ateş and Akın 2014). Green ports resort to proactive development, execution, and follow-up practices with the aim of reducing environmental impacts. Similarly, sustainable port management should be able to develop strategies and activities that respond to the needs of port users to sustain human and natural resource conservation. In this framework, the concepts of green port and sustainable port are interrelated (Dinwoodie et al. 2012). In this context, the green port concept is already considered a part of sustainability, and the way to be a fully sustainable port is to be a green port.

To emphasize the importance of the green port concept, it is key to analyze the environmental impacts originating from the maritime sector and to put forward research that can form a basis for future targets. The study of United Nations (UNs), which proved that global warming due to greenhouse gases (GHG) is a reality for the first time in 1990 by an Intergovernmental Panel on Climate Change (IPCC) formed by the World Meteorological Organization (WMO) and the United Nations Environment Program (UNEP), revealed the necessity of reducing gas emissions (International Maritime Organization n.d.). As a result of understanding the seriousness of the findings of this panel, the Kyoto Protocol, which entered into force in 2005, included provisions to reduce GHG emissions, especially from the aviation and shipping industries (United Nations n.d.). These sectors are treated differently from other greenhouse gas sources. In the GHG studies carried out by the IMO, the greenhouse gas emission rates from maritime transport among the global greenhouse gas sources have been calculated as 3.5% in 2007, 2.6% in 2012, and 2.9% in 2018 (Lee et al. 2009; IMO 2014; Faber et al. 2020). These values were recorded as double the aviation activities for the 2007–2012 period (IMO 2014). Long-term economic and energy-based scenarios set out in the fourth GHG study predicted that maritime transport greenhouse gas emissions, which declined to 90% of 2008 greenhouse gas emissions in 2018, would reach up to 130% in 2050 compared to 2008 (Faber et al. 2020). Globally, the World Health Organization recognizes air pollution as a significant environmental risk and estimates that air pollution, including maritime activities, causes approximately 4.2 million premature deaths (World Health Organization n.d.). In the eco-port initiative carried out by ESPO, air pollution is shown at the top of the environmental priorities of European Union (EU) ports (The European Sea Ports Organisation n.d.).

It is an undeniable fact that port activities have a share in the environmental impacts arising from maritime activities. Ports are not only for maritime activities, but also one of the most important and fundamental elements of the logistics chain. Since ports generally serve in locations close to city centers or populated industrial areas, environmental impacts originating from ports spread to human habitats more quickly and easily. Due to the environmental impacts originating from the ports affect more sensitive areas, the ports have many duties to reduce the environmental impacts, especially the greenhouse gas emissions originating from the ports. Perhaps the most important of these tasks will be to improve the environmental awareness of both port authorities and port users.

Green port projects and programs are carried out to increase the use of renewable energy within the effective operation processes of the ports, regulate their relations with the local community, to further improve environmental and occupational health awareness, and develop competitive port facilities. It is voluntary for ports to participate in green port projects and programs and complete the certification process. Some projects, programs, and award system initiatives implemented in different parts of the world to promote and develop the green port concept are given in Table 9.2.

In order to have a green port awareness, port facilities must first comply with the sectoral criteria determined by the national and international regulations to which the port state is a party and complete the necessary certifications. Generally, ports that want to have the title of green port due to obtain some certificates issued by the international standardization organization. These certificates are as follows (Ministry of Transport and Infrastructure of the Republic of Turkiye Green Port n.d.; Inter-American Committee on Ports n.d.; Organisation and Tools n.d.):

Table 9.2 Green port and shipping projects and programs in the world

Name	Organization	Coverage
Eco-Ports Initiative	ESPO	Ports in Europe
Green Port Project	Republic of Turkey Ministry of Transport and Infrastructure, General Directorate of Maritime Affairs	Ports in Turkey
Green Port Award System (GPAS)	APEC Port Services Network (APSN)	Ports in Asia–Pacific
Green Award	Green Award Foundation	Shipping companies and ships
Maritime Singapore Green Initiative	Maritime and Port Authority of Singapore	Shipping companies and ships
Green Marine	Green Marine Management Corporation	All maritime enterprises in North America

- ISO 9001 Quality Management System (QMS),
- ISO 14001 Environmental Management System (EMS), and
- ISO 18001 Occupational Health and Safety Assessment Series (OHSAS).

The fact that the ports participating in the green port concept themed projects and programs to ensure the quality, environment, and occupational health and safety standards at the national and international level are one of the primary and most important criteria for becoming a green port. In addition, the rules set in international environmental conventions/committees/protocols to which port states are a party is the key motivation for maintaining the green port concept. Some important international conventions/committees/protocols that focus on environmental issues and that can contribute to the development and maintenance of the green port concept are given in Table 9.3.

Although there is no direct reference to port-related environmental impacts in the international conventions/committees/protocols listed in Table 9.3, environmental impacts originating from port activities are also included, albeit indirectly, while evaluating the environmental impacts from both global scale and maritime activities. Therefore, the provisions of the conventions/committees/protocols to which the port state is a party must be implemented by the ports to have the title of green port.

When the environmental priorities of the ports (Table 9.1), projects and award programs for the development of the green port concept (Table 9.2), and international conventions/committees/protocols related to the environment (Table 9.3) are evaluated together, a holistic green port concept that addresses almost all environmental approaches to ports can be designed. Bringing all the components of the port together in a single pot with an environmentalist awareness will make a very important contribution to the creation of the holistic green port concept (Fig. 9.2).

Table 9.3 Some international conventions, committees, and protocols contributing to the green port concept

Convention/Committee/Protocol	Effective year	Parties
Marine Environmental Protection Committee (MPEC)	1974	IMO member states
Convention on the Prevention of Marine Pollution by Dumping of Wastes and Other Matter (London Convention)	1975	87 countries
International Convention on Civil Liability for Oil Pollution Damage (CLC)	1975	136 countries
International Convention for the Safety of Life at Sea (SOLAS)	1980	167 countries
International Convention for the Prevention of Pollution from Ships (MARPOL 73/78)	1983	156 countries
Montreal Protocol	1989	197 countries
Basel Convention	1989	187 countries
United Nations Convention on the Law of the Sea (UNCLOS)	1994	167 countries and EU
Global Program of Action for the Protection of the Marine Environment from Land-based Activities (GPA)	1995	193 countries
International Convention on Oil Pollution Preparedness, Response and Co-operation (OPRC)	1995	112 countries
Helsinki Convention	2000	Baltic countries
International Convention on Civil Liability for Bunker Oil Pollution Damage (BUNKER)	2001	90 countries
Rotterdam Convention	2004	161 countries
Kyoto Protocol	2005	192 countries
International Convention for the Control and Management of Ships' Ballast Water and Sediments (BWM)	2017	86 countries

9.3 Green Port Practices

Although maritime transport is a mode of freight transport with the least environmental impact per cargo unit carried, it is questioned more intensely today because of increasing environmental awareness. This is also mainly due to the slow progress of improvements in reducing environmental impacts of the maritime transport. However, Cullinane and Cullinane (2019) concluded that the combination of regulations and technological innovation with the support of ports could have significant potential to reduce environmental impacts. In this context, although having a green port label is completely voluntary, it can be said that environmental-awareness ports can play a pivotal role in achieving the goal of reducing the environmental impact of maritime transport.

Fig. 9.2 A holistic green port concept

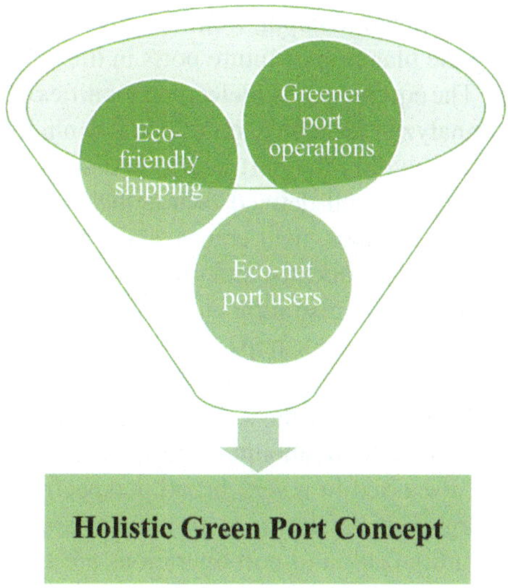

While in port, the total emission amount of ships is measured less than when cruising. However, due to the urban location of most ports, urban residents are directly exposed to pollutants sourcing from ships in the port area (Styhre and Winnes 2019). Therefore, while the ports reduce the environmental impacts arising from their own operations, they should also apply measures to reduce the environmental impacts caused by the ships calling at the port and the land-based stakeholders using the port, so that it can be considered as an eco-friendly port in a holistic approach. In this sense, it would be more appropriate using a holistic green port approach to evaluate practical green port applications in three categories.

- Green port practices to reduce environmental impacts of port operations
- Green port practices to reduce environmental impacts of ships
- Green port practices to reduce environmental impacts of shore-side port users.

9.3.1 Green Port Practices to Reduce Environmental Impacts of Port Operations

The environmental priority of ports with a green port approach is to increase the air quality around the port (The European Sea Ports Organisation n.d.). Accordingly, some ports define themselves as innovative ecosystems for a zero-emission society in the symbiosis of ports and port cities (Jansen 2020). To create a zero-emission port and port city symbiosis, ports have important duties in terms of port operations. The port of Rotterdam is making a big breakthrough in this regard, with the target of a virtually emission-free port for 2050. Nowadays, it is primarily essential to minimize emissions (Davis et al. 2018). Afterward, to ensure sustainability, many issues should be examined in depth, from the development of a green port management approach and the planning of future ports in this direction.

The equipment, vehicles, and facilities of a port that emits greenhouse gases should be analyzed in depth to minimize greenhouse gas emissions. In this way, the emission sources within the port facility are determined, and the emitters that reduce the air quality are well-defined. While the prominent equipment and vehicles in the ports are quay cranes, field cranes, stacking-storage cranes and vehicles, terminal trucks and chassis, forklifts, and other port field vehicles, the facilities can be identified as the main administration building, harbormaster office, security and customs offices, maintenance workshops, terminal operation buildings, parking lots, social facilities, etc.

Depending on the variety of equipment, vehicles and facilities in the ports and the volume and amount of port activities, practical green port applications in ports are discussed in a very broad perspective. However, transition to renewable green energy can be shown as prominent applications among them. Electrification and decarbonization of port operations are keyways to mitigate port-related greenhouse gas impacts. Hereby, studies on how to decarbonize port operations are of great importance (Psaraftis 2019; Samadi et al. 2016; Pastra et al. 2021). Additionally, clean electricity cannot be reached in electricity production from coal or biomass (Faaij et al. 1998; Diji 2013; Loução et al. 2019). However, electricity generation with renewable energy sources significantly reduces the carbon intensity (Robyns et al. 2012; Al-Mansour et al. 2014; Verhaegen et al. 2007). Many activities hosted by ports can use electricity, and these activities can be decarbonized through electrification. It is possible to ensure the electrification in the following port activities: operation of handling equipment and port logistics vehicles, bunkering operations, cold or hot storage operations (cooling of containers, or heating of chemical tanks), pilotage, and tugboat services.

In addition to electrifying port-related activities as a green transition toward decarbonization, the ports are also aiming to grow sustainability by investing in innovative technologies minimizing the electricity consumption in existing electrical equipment and facilities. At this point, the fact that the electric motors used with electrification are more efficient; safer and less maintenance-cost needed operations can be observed as the benefit of innovative technologies to the green port approach (Hashemnia and Asaei 2008; Ren et al. 2009). Another example in this context is that the Port of Rotterdam, the largest port in Europe, which uses the light-emitting diode (LED) lighting as an innovative technology in the entire port in 2022 (Port of Rotterdam n.d.).

To reduce the environmental impacts of port operations, it is also possible to see the following practices regarding the green port approach. Some of those:

- Considering environmental priorities in purchasing new equipment, vehicles, and materials (less air pollutant, energy efficient, less noisy, etc.),
- Ensuring energy recycling in port equipment and vehicles,
- Environmentally friendly offices and buildings that produce their own energy or minimize the energy use.

Port planning in line with green port policies is also important in reducing the environmental impacts of port operations. Effective administrative, operational, and technical planning is required for port planning to support the green port approach. At this point, for the administrative structure, firstly the required number of personnel with the necessary qualifications should be employed (Thai 2012). Environmental awareness can be instilled by providing qualified and periodic orientation and in-house trainings for the employed staff (Vaio and Varriale 2018). The use of software programs developed at the point of effective operational planning significantly increases operational efficiency. In this way, efficient planning can be made by minimizing the waiting times of the equipment and vehicles used in the operation or by eliminating insufficient or idle capacity problems (Gosasang et al. 2018; Bichou 2014). In terms of technical port planning, it is important that the port is equipped with eco-friendly equipment and vehicles suitable for its operational characteristics, and these equipment and vehicles are assigned to jobs that comply with their technical specifications (Rose et al. 2022).

As a result of green port planning, the ports are expected to take measures in accordance with national and international legal regulations regarding waste and pollution factors. Some of these measures can be categorized as follows:

- Measures regarding the collection and disposal of domestic solid waste
- Measures regarding the collection, separation, control, or disposal of packaging waste
- Precautions for the collection and disposal of waste batteries and accumulators
- Measures regarding hazardous waste
- Measures regarding the control of waste from ships
- Measures to prevent marine pollution (such as water treatment plant-Evyapport, storm water management program—Port of San Diego)
- Measures to reduce the effects of chemicals used in cleaning the port area, equipment, and vehicles.

9.3.2 Green Port Practices to Reduce Environmental Impacts of Ships

The release of SO_x, NO_x, and particulate material (PM), which adversely affects human health, from the ships calling at the ports raises the concerns due to the closeness of the ports to the cities. The most serious studies on reducing these emissions from ships are carried out and supported by the IMO (Lee et al. 2009; IMO 2014; Faber et al. 2020). In this direction, the IMO has set out a target of reducing greenhouse gas emissions from ships by 50% in 2050 compared to 2008 (International Maritime Organization n.d.).

It is very important to reduce or eliminate emissions from ships in ports, both in terms of creating clean shipping in terms of the IMO targets and developing the green port approach. Different methods can be used to reduce the emissions of ships in ports. One of them is to change the type of fuel used on ships. What stands out at this point is the transition to fuels with less greenhouse gas emissions, which are defined as an alternative to conventional fuels (Kolwzan and Narewski 2012; Eide et al. 2013). These alternative fuel types can be grouped as liquefied natural gas (LNG), liquefied petroleum gas (LPG), hydrogen-based fuels, electricity, ammonia, and methanol.

Both shipowners and ports have important duties in switching the conventional fuel type to less emission one. It is one of the most basic actions for shipowners to act primarily to reduce the use of heavy fuel oil (HFO). Current international regulations on this subject already impose restrictions on switching to marine gas oil (MGO) or using scrubbers in special areas such as ((Emission control areas (ECAs) and sulfur emission control areas (SECAs)) (Tran et al. 2017; Zetterdahl et al. 2016). In addition to the ECA and SECA areas, ships are commonly switching to MGO or marine diesel oil in their auxiliary machinery in ports, reducing the amount of emissions (Kose et al. 2022; Chen et al. 2021).

LNG is seen as the most promising alternative fuel in today's conditions. However, according to the data announced by the World Port Climate Initiative (WPCI 2018), the number of ships in the world using LNG (excluding LNG carriers) is very small (World Port Sustainability Program n.d.). The underlying reason for this is that there are insufficient LNG supply facilities, and supply problems will arise if LNG becomes widespread (Wang and Notteboom 2014). It can be said that the duty of ports to spread LNG usage is to make suitable LNG supply facilities investments in port facilities (Styhre et al. 2017). The use of LNG is attractive not only in the open sea navigation, but also in reducing emissions in port environment and operation area.

Other options being considered to reduce emissions are hydrogen and electricity. However, since hydrogen and electricity are still produced using non-green methods like coal, there is a concern about eliminating greenhouse gas emissions (Midilli et al. 2021). However, the use of renewable energy sources in hydrogen and electricity production increases environmental attractiveness. Ports can help accelerate the global decarbonization approach while reducing shipping emissions and implementing green port plans by adopting hydrogen transition technologies. A database of projects carried out in ports that adopt hydrogen as a clean energy source has been created by the International Association of Ports and Harbors (IAPHs). Accordingly, the projects carried out in ports on hydrogen technologies in the world are given in Table 9.4 (World Port Sustainability Program n.d.).

The technology of supplying electricity to ships in port during their stay at the port, called cold ironing or onshore power supply (OPS) or shore-side electricity (SSE), is also a method to reduce the emissions of ships while in port (Zis 2019). However, it is obvious that if the electricity provided is from renewable sources, it will be effective in terms of reducing emissions. Otherwise, while reducing emissions only in the port area, emissions will still be significant in the location where electricity is generated. Vaishnav et al. (2016) and Winkel et al. (2016), in their studies on this subject, determined that with cold ironing, significant emission reductions can be achieved, and contributions can be made to the environmental, economic, and health of the people of the region. However, the number of ports with cold ironing

Table 9.4 Hydrogen projects carried by world ports. (*Source* Adapted from WPSP (World Port Sustainability Program n.d.))

Country	Port name	Project name	Project date
New Zealand	Ports of Auckland	GHG Emission Reduction Pathway	2018
Japan	Port of Kobe	Environmental Measures in Reclamation Projects	2018
United States	Port of Los Angeles	Zero Emissions Pathway Technology Demonstrations	2019
France	Port of Marseille	HyAMMED	2019
France	Port of Marseille	Jupiter 1000	2019
United States	Port of Long Beach	C-PORT Zero Emissions Demonstration Project	2019
Germany	Niedersachsen Ports	WASh2Emden project	2019
Spain	Port of Valencia	H2Ports/Fuel Cells and Hydrogen in Ports	2019
France	Port of Marseille	Green mobile energy for reefer containers	2020
France	Port of Marseille	Energy recovery from cruise ships' wastewater	2020
United Kingdom	Port of London Authority	Hydrogen Highway	2021
Netherlands	Port of Amsterdam	Multi Fuel Port: Spatial Safety	2021
Sweden	Port of Gothenburg	Hydrogen production facility and filling station	2021
Japan	Port of Yokohama	Hydrogen Supply Chain Joint Study	2021
Sweden	Port of Gothenburg	Tranzero Initiative	2021
Netherlands	Port of Amsterdam	H2SHIPS project	2021
Austria	Port of Vienna	H2 meets H2O	2022

system is very small. The main obstacles are installation costs and the ships' lack of connection points in international standards that can be integrated into the system (Innes and Monios 2018). However, significant savings can be achieved with the legal regulations on this subject and the provision of international standards.

The biofuel alternative, on the other hand, is not expected to be very common in maritime transportation due to production methods and intensive land use (Kesieme et al. 2019). Navigation with onboard battery installation, which is shown as another alternative, can only be foreseen for short distance trade with current battery technology (Haxhiu et al. 2021). The energy requirement and ship sailing from wind and solar power, which are seen as the cleanest alternative energy sources, are still seen as a niche alternative, and its sustainability is a matter of debate (Friebe et al. 2017).

It is valuable that shipowners tend to apply the necessary technical improvements and innovative technologies on their ships to use cleaner fuel type in compliance with legal regulations. Cleaner maritime transport, which will be created by shipowners who have the tendency and sensitivity to be greener and take action, will make significant contributions to the green port approach. In addition to all these greener fuel transition predictions, significant reductions in fuel consumption and greenhouse gas emissions can be recorded along with structural improvements on ships (improved hull design, more efficient engines, etc.) (Palomba et al. 2022) and operational improvements on voyages (slow steaming, increased operational efficiency due to on time arrival and departure, weather routing, etc.) (Viktorelius and Lundh 2019; Beşikçi et al. 2016).

Award programs are carried out to encourage ships, ports, and maritime companies to carry out maritime activities in an eco-friendlier manner (see Table 9.2). Ships, ports, and shipping companies that have participated in such programs and received a green award or green certificate have the opportunity to receive many incentives by program providers. Some of these incentives can be listed as discount on the port dues, discount on a service or a product, special extra service, or product and promotion. In this way, if the incentive provider is a port, a green-certified ship can receive a discount for the services when the ship calls the port. For example, the port of Rotterdam offers discounts for LNG carriers, chemical/gas tankers, and crude oil/product tankers with Green Award certification (Port of Rotterdam n.d.).

9.3.3 Green Port Practices to Reduce Environmental Impacts of Shore-Side Port Users

There is an increasing interest in the green port concept and a growing literature with research being conducted. However, studies on associating the environmental effects originating from the hinterland with the ports are limited, and there are only a few of them (Lam and Notteboom 2014; Acciaro et al. 2014). The ports are hubs and intermodal transit points for global freight transport systems. Therefore, for hinterland logistics, ports have the potential to lead the decision mechanisms of both local and global logistics providers with their infrastructure facilities (road, rail, and inland shipping connections and agreements) and the strategies they implement. It can be said that the role of ports in hinterland emissions is partially responsible, if not as much as in their own operations. Because providing sufficient hinterland connection, alternatives can be considered within this partial responsibility.

Emissions originating from port users are not calculated as emissions originating from the port, and these are calculated and recorded as externalities. However, hinterland transportation is the continuation of a sea transportation system by changing the mode at the port. And aiming for a total environmental improvement requires interaction, harmony, and cooperation between ports and their users. Bergqvist and Egels-Zandén (2012) state that ports are key actors to increase the environmental efficiency of hinterland transport and show that they can easily promote an environmentally friendly hinterland with a strategy such as a green port due.

A few examples can be cited that encourage the greening of port users' activities. Some of these examples are the PierPASS program at the ports of Los Angeles and Long Beach, the port terminal concession agreements at the port of Rotterdam, and the differentiated port dues system (Bergqvist and Egels-Zandén 2012; Giuliano and O'Brien 2008; Berg and Langen 2014). Bergqvist et al. (2015) determined that internalization of externalities, road pricing, modal split quota, and additional port charges are used by ports to encourage port users and take environmental measures to reduce the environmental impacts of hinterland transport. Aregall et al. (2018) found that only 76 ports in the world have implemented a strategy to reduce the environmental impacts of the hinterland in developing green ports. The most widely implemented strategy is to reduce air pollution, which is one of the environmental priorities of ports. This is followed by congestion, mode shift, and noise pollution, respectively.

9.4 Conclusions

This chapter deals with the green port projects, programs, and award incentives carried out to enable the development of the green port concept. It also evaluates the practices that can be implemented in ports because of the environmental awareness that is tried to be gained through the incentives provided. Thus, as a result of the current situation assessment, how green port practices can be done more, and better is discussed in depth.

Significant financial budgets are required for the establishment of infrastructures related to the environmental practices of ports. The truth is that ports are commercial businesses, and they don't want to spend money unless it's necessary. Therefore, ports are expected to be supported by the government with incentives during the transition to "green port" practices. This ensures both the implementation of currently known best practices and faster adoption of new technology. Because as long as having the title of green port depends on voluntariness, ports will not want to spend money unless they have environmental awareness. However, thanks to government incentives, the green port concept will be encouraged to be brought into the ports. For example, with tariff regulations to promote electrification, the transition to electrical equipment and vehicles can be accelerated, and the use of clean energy can be increased. Or state grants can be provided for infrastructure investments for ports that want to produce electricity and hydrogen.

Reducing and eliminating emissions from port operations alone is not enough for the green port concept. In addition, the green port concept can be developed, and its sustainability can be possible with the environmental awareness of all stakeholders associated with the port. Thus, port stakeholders, which will be motivated by ports have a pivotal role, can reduce their own emission resources, and a greener and more sustainable maritime transport can be built with innovative technologies. While greater sharing of best practices and further technological development are needed, and although there is enough information already available to make major advances in reducing environmental externalities, slow adoption in the marine industry is prolonging the transition to the green port concept. Although accelerating adaptation is dependent on government incentives, this sluggishness will continue as long as the environmental awareness of states, ports, and port stakeholders is not improved. When additional costs arise, the motivation to act will disappear.

This chapter evaluates the green port concept for the decarbonization of the maritime industry by considering green port practices from three perspectives. They are (i) port operations, (ii) shipping, and (iii) hinterland activities. There are still some limitations in the improvement, development, and implementation of the green port concept, so further studies can be approached from the following directions. First, cost reduction studies can be carried out to increase the applicability of green port practices in ports. Secondly, considering different country and region characteristics, some specific green port practices can be developed for a better green port concept for a particular port or port area. Finally, the concept of green port can be included in national or international legislation to implement and spread the best green port practices at a certain standard.

As a result, in the process of becoming a green port, of course, there will be expectations from state authorities. However, port management, employees, and stakeholders should also be respectful to the natural environment, not worry about their families and children living in the port area and be conscious and carry out their activities with this awareness. It is an important criterion to prepare training programs with the necessary qualifications and qualifications to create total environmental awareness and sensitivity. To adopt the green port concept in all ports, it is also an option to take legal regulation and fine enforcement actions as well as environmental awareness.

References

Acciaro M, Vanelslander T, Sys C, Ferrari C, Roumboutsos A, Giuliano G, Lam JSL, Kapros S (2014) Environmental sustainability in seaports: a framework for successful innovation. Marit Policy Manag 41(5):480–500. https://doi.org/10.1080/03088839.2014.932926

Al-Mansour F, Sucic B, Pusnik M (2014) Challenges and prospects of electricity production from renewable energy sources in Slovenia. Energy 77:73–81. https://doi.org/10.1016/j.energy.2014.04.106

Aregall MG, Bergqvist R, Monios J (2018) A global review of the hinterland dimension of green port strategies. Transp Res Part d: Transp Environ 59:23–34. https://doi.org/10.1016/j.trd.2017.12.013

Ateş A, Akın M (2014) Türkiye'de Yeşil Liman Kavramı ve Yasal Çerçevesi. Akademik Platform. (In Turkish)

Axel BN (ed) (2011) Environmental impacts of international shipping the role of ports: the role of ports, vol 2011. OECD Publishing

Bergqvist R, Egels-Zandén N (2012) Green port dues—The case of hinterland transport. Res Transp Bus Manag 5:85–91. https://doi.org/10.1016/j.rtbm.2012.10.002

Bergqvist R, Macharis C, Meers D, Woxenius J (2015) Making hinterland transport more sustainable a multi actor multi criteria analysis. Res Transp Bus Manag 14:80–89. https://doi.org/10.1016/j.rtbm.2014.10.009

Beşikçi EB, Kececi T, Arslan O, Turan O (2016) An application of fuzzy-AHP to ship operational energy efficiency measures. Ocean Eng 121:392–402. https://doi.org/10.1016/j.oceaneng.2016.05.031

Bichou K (2014) Port operations, planning and logistics. CRC Press

Chen S, Meng Q, Jia P, Kuang H (2021) An operational-mode-based method for estimating ship emissions in port waters. Transp Res Part d: Transp Environ 101:103080. https://doi.org/10.1016/j.trd.2021.103080

Cullinane K, Cullinane S (2019) Policy on reducing shipping emissions: implications for “green ports”. Green Ports, pp35–62. https://doi.org/10.1016/B978-0-12-814054-3.00003-7

Davis SJ, Lewis NS, Shaner M, Aggarwal S, Arent D, Azevedo IL, Benson SM, Bradley T, Brouwer J, Chiang YM, Clack CT (2018) Net-zero emissions energy systems. Science 360(6396):p.eaas9793. https://doi.org/10.1126/science.aas9793

Di Vaio A, Varriale L (2018) Management innovation for environmental sustainability in seaports: Managerial accounting instruments and training for competitive green ports beyond the regulations. Sustainability 10(3):783. https://doi.org/10.3390/su10030783

Diji CJ (2013) Electricity production from biomass in Nigeria: options, prospects and challenges. Adv Mater Res 824:444–450. https://doi.org/10.4028/www.scientific.net/AMR.824.444

Dinwoodie J, Tuck S, Knowles H, Benhin J, Sansom M (2012) Sustainable development of maritime operations in ports. Bus Strateg Environ 21(2):111–126

Du K, Monios J, Wang Y (2019) Green port strategies in China. In: Green ports, pp 211–229. Elsevier

Eide MS, Chryssakis C, Endresen Ø (2013) CO_2 abatement potential towards 2050 for shipping, including alternative fuels. Carbon Manag 4(3):275–289. https://doi.org/10.4155/cmt.13.27

Faaij A, Meuleman B, Turkenburg W, vAN Wijk AD, Bauen A, Rosillo-Calle F, Hall D (1998) Externalities of biomass based electricity production compared with power generation from coal in the Netherlands. Biomass Bioenerg 14(2):125–147. https://doi.org/10.1016/S0961-9534(97)10005-8

Faber J, Hanayama S, Zhang S, Pereda P, Comer B, Hauerhof E, van der Loeff WS, Smith T, Zhang Y, Kosaka H, Adachi M, Bonello J, Galbraith C, Gong Z, Hirata K, Hummels D, Kleijn A, Lee DS, Liu Y, Lucchesi A, Mao X, Muraoka E, Osipova L, Qian H, Rutherford D, Fuente SS, Yuan H, Perico CV, Wu L, Sun D, Yoo DH, Xing H (2020) Fourth IMO greenhouse gas study

Friebe A, Olsson M, Le Gallic M, Springett JL, Dahl K, Waller M (2017) A marine research ASV utilizing wind and solar power. In: OCEANS 2017-Aberdeen, pp 1–7. IEEE. https://doi.org/10.1109/OCEANSE.2017.8084648

Giuliano G, O’Brien T (2008) Extended gate operations at the ports of Los Angeles and Long Beach: a preliminary assessment. Marit Policy Manag 35(2):215–235. https://doi.org/10.1080/03088830801956854

Gosasang V, Yip TL, Chandraprakaikul W (2018) Long-term container throughput forecast and equipment planning: the case of Bangkok Port. Maritime Bus Rev. https://doi.org/10.1108/MABR-07-2017-0019

Hashemnia N, Asaei B (2008) Comparative study of using different electric motors in the electric vehicles. In: 2008 18th international conference on electrical machines, pp 1–5. IEEE. https://doi.org/10.1109/ICELMACH.2008.4800157

Haxhiu A, Abdelhakim A, Kanerva S, Bogen J (2021) Electric power integration schemes of the hybrid fuel cells and batteries-fed marine vessels–an overview. IEEE Trans Transp Electrification. https://doi.org/10.1109/TTE.2021.3126100

IMO, Third IMO (2014) Greenhouse Gas Study 2014, executive summary and final report. Int Maritime Orga, London, 280

Innes A, Monios J (2018) Identifying the unique challenges of installing cold ironing at small and medium ports–The case of Aberdeen. Transp Res Part d: Transp Environ 62:298–313. https://doi.org/10.1016/j.trd.2018.02.004
Inter-American Committee on Ports. Green port guide translation. https://portalcip.org/
International Maritime Organization. Historic background. https://www.imo.org/en/OurWork/Environment/Pages/Historic%20Background%20GHG.aspx
International Maritime Organization. Initial IMO GHG strategy. https://www.imo.org/en/MediaCentre/HotTopics/Pages/Reducing-greenhouse-gas-emissions-from-ships.aspx.
Jansen M (2020) Port innovation ecosystem, a symbiosis of capital
Kesieme U, Pazouki K, Murphy A, Chrysanthou A (2019) Biofuel as an alternative shipping fuel: technological, environmental and economic assessment. Sustain Energy Fuels 3(4):899–909. https://doi.org/10.1039/C8SE00466H
Kolwzan K, Narewski M (2012) Alternative fuels for marine applications. Latvian J Chem 51(4):398. https://doi.org/10.2478/v10161-012-0024-9
Kose S, Sekban DM, Ozkok M (2022) Determination of port-induced exhaust gas emission amounts and investigation of environmental impact by creating emission maps: Sample of Trabzon port. Int J Sustain Transp 16(3):258–268. https://doi.org/10.1080/15568318.2020.1871130
Lam JSL, Notteboom T (2014) The greening of ports: a comparison of port management tools used by leading ports in Asia and Europe. Transp Rev 34(2):169–189. https://doi.org/10.1080/01441647.2014.891162
Lee D, Lindstad H, Markowska AZ, Mjelde A, Nelissen D, Nilsen J, Pålsson C, Winebrake JJ, Wu W, Yoshida K (2009) Second IMO greenhouse gas study
Loução PO, Ribau JP, Ferreira AF (2019) Life cycle and decision analysis of electricity production from biomass–Portugal case study. Renew Sustain Energy Rev 108:452–480. https://doi.org/10.1016/j.rser.2019.03.063
Luo M, Yip TL (2013) Ports and the environment. Marit Policy Manag 40(5):401–403
Midilli A, Kucuk H, Topal ME, Akbulut U, Dincer I (2021) A comprehensive review on hydrogen production from coal gasification: challenges and opportunities. Int J Hydrogen Energy 46(50):25385–25412. https://doi.org/10.1016/j.ijhydene.2021.05.088
Ministry of Transport and Infrastructure of the Republic of Turkiye Green Port. https://denizcilik.uab.gov.tr/yesil-liman
Munim ZH, Sornn-Friese H, Dushenko M (2020) Identifying the appropriate governance model for green port management: applying analytic network process and best-worst methods to ports in the Indian Ocean Rim. J Clean Prod 268:122156
Oniszczuk-Jastrząbek A, Pawłowska B, Czermański E (2018) polish sea ports and the green port concept. In: SHS web of conferences, Vol 57, p 01023. EDP Sciences
Palomba G, Scattareggia Marchese S, Crupi V, Garbatov Y (2022) Cost, energy efficiency and carbon footprint analysis of hybrid light-weight bulk carrier. J Marine Sci Eng 10(7):957. https://doi.org/10.3390/jmse10070957
Pastra A, Doelle M, Johansson T (2021) The shipping sector and ports as central actors in the Decarbonization effort: a case study of China. Asia-Pacific J Ocean Law Policy 6(2):247–270
Port of Rotterdam. Green Award Discount. https://www.portofrotterdam.com/en/sea-shipping/seaport-dues/green-award-discount
Port of Rotterdam. Sustainable port. https://www.portofrotterdam.com/en/building-port/sustainable-port
Psaraftis HN (2019) Decarbonization of maritime transport: to be or not to be? Maritime Econ Logist 21(3):353–371

Puig M, Wooldridge C, Michail A, Darbra RM (2015) Current status and trends of the environmental performance in European ports. Environ Sci Policy 48:57–66
Ren Q, Crolla DA, Morris A (2009) Effect of transmission design on electric vehicle (EV) performance. In: 2009 IEEE vehicle power and propulsion conference, pp 1260–1265. IEEE. https://doi.org/10.1109/VPPC.2009.5289707
Robyns B, Davigny A, François B, Henneton A, Sprooten J (2012) Electricity production from renewable energies, pp 75–146. Wiley
Rose H, Lange AK, Hinckeldeyn J, Jahn C, Kreutzfeldt J (2022) Investigating the requirements of automated vehicles for port-internal logistics of containers. In: International conference on dynamics in logistics, pp 179–190. Springer, Cham
Samadi S, Lechtenböhmer S, Schneider C, Arnold K, Fischedick M, Schüwer D, Pastowski A (2016) Decarbonization pathways for the industrial cluster of the port of Rotterdam. Wuppertal, Germany: Wuppertal Institute for Climate, Environment and Energy
Styhre L, Winnes H, Black J, Lee J, Le-Griffin H (2017) Greenhouse gas emissions from ships in ports–case studies in four continents. Transp Res Part d: Transp Environ 54:212–224. https://doi.org/10.1016/j.trd.2017.04.033
Styhre L, Winnes H (2019) Emissions from ships in ports. In: Green ports, pp 109–124. Elsevier. https://doi.org/10.1016/B978-0-12-814054-3.00006-2
Thai VV (2012) Competencies required by port personnel in the new era: conceptual framework and case study. IJSTL-Int J Shipping Transp Logist 4(1):49. https://doi.org/10.1504/IJSTL.2012.044135
The European Sea Ports Organisation. EcoPorts Tools. Port environmental review system (PERS). https://www.ecoports.com/pers
The European Sea Ports Organisation. EcoPorts. https://www.ecoports.com
The United Nations Conference on Trade and Development. Review of Maritime Transport 2021. https://unctad.org/webflyer/review-maritime-transport-2021
Tran VD, Le AT, Hoang AT (2017) Methods of operating the marine engines by ultra-low sulfur fuel to aiming to satisfy MARPOLAnnex VI. Adv Nat Appl Sci 11(12):34–41
United Nations. Climate change. The Kyoto Protocol. https://unfccc.int/kyoto_protocol
Vaishnav P, Fischbeck PS, Morgan MG, Corbett JJ (2016) Shore power for vessels calling at US ports: benefits and costs. Environ Sci Technol 50(3):1102–1110. https://doi.org/10.1021/acs.est.5b04860
Van den Berg R, De Langen PW (2014) An exploratory analysis of the effects of modal split obligations in terminal concession contracts. Int J Shipping Transp Logist 6(6):571–592
Verhaegen K, Meeus L, Delvaux B, Belmans R (2007) Electricity produced from renewable energy sources—What target are we aiming for? Energy Policy 35(11):5576–5584. https://doi.org/10.1016/j.enpol.2007.06.004
VERİFAVİA. Local Green Port Programmes. https://www.verifavia.com/greenhouse-gas-verification/program-local-green-port-programmes-350.php
Viktorelius M, Lundh M (2019) Energy efficiency at sea: an activity theoretical perspective on operational energy efficiency in maritime transport. Energy Res Soc Sci 52:1–9. https://doi.org/10.1016/j.erss.2019.01.021
Wang S, Notteboom T (2014) The adoption of liquefied natural gas as a ship fuel: a systematic review of perspectives and challenges. Transp Rev 34(6):749–774. https://doi.org/10.1080/01441647.2014.981884
Winkel R, Weddige U, Johnsen D, Hoen V, Papaefthimiou S (2016) Shore side electricity in Europe: potential and environmental benefits. Energy Policy 88:584–593. https://doi.org/10.1016/j.enpol.2015.07.013
World Health Organization. Ambient (outdoor) air pollution. https://www.who.int/en/news-room/fact-sheets/detail/ambient-(outdoor)-air-quality-and-health
World Port Sustainability Program. IAPH-Onshore power supply. https://sustainableworldports.org/project/iaph-onshore-power-supply/

World Port Sustainability Program. Global Ports Hydrogen coalition. https://sustainableworldports.org/global-ports-hydrogen-coalition/?cpage=3&fuel=hydrogen
Zetterdahl M, Moldanova J, Pei X, Pathak RK, Demirdjian B (2016) Impact of the 0.1% fuel sulfur content limit in SECA on particle and gaseous emissions from marine vessels. Atmos Environ 145:338–345. https://doi.org/10.1016/j.atmosenv.2016.09.022
Zis TP (2019) Prospects of cold ironing as an emissions reduction option. Transp Res Part a: Policy Pract 119:82–95. https://doi.org/10.1016/j.tra.2018.11.003

Part V
Operational Measures

Chapter 10
Energy Efficiency and Management Onboard Ships

M. Ziya Sogut and Suleyman Ozkaynak

Abstract Maritime transport, which makes up 90% of the world's transport, has improved its responsibilities in combating global climate change with its 2050 commitments within the scope of COP 26. In this context, ships are developing their studies on alternative solutions to reduce fossil fuel consumption. However, the institutional lack of energy management and sustainable energy efficiency for ships is an important problem. In this study, first of all, the energy management framework was developed, and the energy efficiency potential was evaluated by performing a gap analysis for a tanker ship. Basic indicators for the manageability of energy were defined and compared. While the energy efficiency potential was 10.17% in the study, the thermal performance average was found to be 36.48%. At the end of the study, the effect of energy management was evaluated, and suggestions were developed for sustainable energy management processes.

Keywords Tanker ship · Energy management · Efficiency · Environmental · Sustainability

10.1 Introduction

The causes of global climate change have increased in importance, with energy management and efficiency being the key issues for institutional structures. Considering the emission effect, environmental problems caused by fossil-based consumption and manageability in the maritime sector, as in all sectors, were taken into account. As a matter of fact, as a sectoral awareness, in 2018, the International Maritime Organization (IMO) decided to reduce the total emissions from the international maritime sector by half of 2008 values by the middle of the twenty-first century (IMO 2018). In addition, reducing fuel-related sulfur emissions from 3.50%

M. Z. Sogut (✉) · S. Ozkaynak
Piri Reis University, Maritime Faculty, Istanbul, Türkiye
e-mail: mzsogut@gmail.com; mzsogut@pirireis.edu.tr

S. Ozkaynak
e-mail: sozkaynak@pirireis.edu.tr

B. Zincir et al. (eds.), *Decarbonization of Maritime Transport*, Energy, Environment, and Sustainability, https://doi.org/10.1007/978-981-99-1677-1_10

m/m to 0.50% m/m outside the control areas (ECA) on ships again shows the solution by IMO (Campara et al. 2018). The NOx content in the exhaust gas of an engine, which is another emission factor, cannot exceed 17.0 g/kW h according to MARPOL Annex IV. In addition to exhaust gas cleaning systems, efficient energy management ensures that emission rates are kept within permissible limits Seddiek and Elgohary (2014). All these measures taken by the IMO show that the attention of policy-makers in the industry is on pollution from energy use. In addition, bunker fuel prices are almost 75% higher than 80–90s prices and are still increasing in the long term (IEA 2019). Today, this fuel price accounts for 43% to 67% of a ship's total operating cost (Kalli et al. 2009). Finally, the collection of fuel consumption, CO_2 emission measurements, speed optimization, and propulsion system maintenance data by authorized onboard personnel demonstrates that energy efficiency awareness has grown (MEPC 2016). Accordingly, the concepts of energy management and efficiency for industry and ships come to the fore.

Considering the sectoral frameworks, energy management is a concept that has been carried out in many studies that focus on energy efficiency. This process, which is basically to reduce fossil fuels, has developed as a goal with a multifaceted effect in the maritime sector, not only energy efficiency, but also environmental impact and economic solutions. In this respect, it is a necessary development with multifaceted regulations, especially IMO.

The focal points of sectoral energy efficiency in ships are trim, speed, paint, and propellers. In this framework, ships have developed some responsibilities, including the design process of the Ships Energy Efficiency Management Plan (SEEMP) on which it is based (MEPC 2016). Especially with the process, many elements that are sectoral requirements on ships cause excessive fuel use in engine power during the voyages. Layer formations caused by marine-based surface pollution can also be seen as an effect in this direction. These conditions may be the cause of inefficiency of up to 10% in terms of total fuel consumption on ships (Adland et al. 2018a, b). One of the industry's focuses is trim control. Depending on the load conditions, trim control and management is an important parameter in terms of energy efficiency. This effect is a point of attention by ship management today and is controlled by processes in energy efficiency, especially traditional means.

Energy efficiency on ships is handled with some indicators defined by the IMO regulations. These indicators, which were developed as a measure of efficiency in the design and operation of ships, are also the basic indicators from the institutional point of view. The Energy Efficiency Design Index (EEDI) and Energy Efficiency Operational Indicator (EEOI) values are evaluated to understand the environmental impacts of the ship's energy consumption. This shows that efficiency is not only based on fuel consumption and power obtained, but can also be calculated in terms of distance traveled, available capacity, cargo carried, ship speed, etc. (Adland et al. 2018a, b). The EEDI is basically an indicator that defines the CO_2 emission potential per mile for the new ship. Its calculation is based on assumptions regarding the installed power on board and the specific fuel consumption of the engines (in g/kWh) compared to the available capacity defined in MEPC.1/Circ.681. EEOI as an indicator that provides information about the efficiency of the ship during voyages shows the CO_2 emissions per nautical mile by a calculation based on a ship's fuel consumption and the transport business, such as the amount of cargo carried or passengers carried (IMO 2009). Unlike EEDI, the EEOI measures the actual efficiency of the ship in operation and can be used for hull and propeller cleanliness, slow steaming, improved voyage planning, etc., can measure the effects of any change, such as in short, EEDI and EEOI calculations show CO_2 emissions per ship use. When energy efficiency is monitored with EEOI perspectives, energy management will be examined numerically.

Ships are under control with SEEMP in standard applications. As in many industries, the ISO 50001 Energy Management System for energy management on ships provides usable guidelines to be a functional energy management system. ISO 50001 provides steps to be used as a corporate model to make energy efficiency continuous and sustainable. Certificates are issued by third-party certification bodies. Audits can be carried out on a regular basis to ensure that the management system is continually being improved. Certification requires an operating company or ship to develop an energy policy, use data to achieve objectives, measure policy effectiveness, and make continuous improvements to policy through a system. The system requires an action plan that specifies exactly how the organization's goals will be achieved. ISO 50001 uses the Plan, Do, Check, Act (PDCA) cycle. In the planning phase, the organization sets goals and targets using existing energy efficiency measures. During the Do phase, the organization implements actions to improve energy efficiency. In the control phase, the organization measures and evaluates its energy performance and compares the results with baseline values. In the action phase, the organization decides what changes need to be made to improve its energy performance. The cycle repeats itself with a new planning phase. Thanks to ISO 50001, the audit of independent organizations other than the classification societies authorized by ISO and other organizations makes the energy management of operating companies and ships more controllable and understandable.

Another outstanding feature of the ISO 50001 energy management standard is that different energy management systems such as SEEMP or ISM can be combined and integrated with ISO 50001 (Knorring et al. 2012). Today, literature examples also present different perspectives on these management models. Seddiek and Elgohary (2014) discussed different methods that can be used to reduce emissions, such as fuel saving strategies. Kalli et al. (2009) This shows that the importance of energy management has increased due to rising fuel prices and solutions for the situation. Baldia et al. (2014) studied the energy system of a chemical tanker and analyzed its energy use. Rehmatulla and Smith (2015) investigated the barriers to energy efficiency in the shipping industry. On the other hand, Georgopoulou et al. (2021) and Adland et al. (2018a, b) examined the effect of the ship's structural features on energy efficiency and showed that the maintenance process effect of the ship can increase the energy efficiency by around 8%. Perera and Mo (2016) analyzed energy efficiency from an emission perspective, gave an overview of energy efficiency measures based on emission control, and comparatively examined EEDI, EEOI, and SEEMP. In this study, ship energy management proposes a framework for energy management and efficiency assessment. It is shown by proposing a model whether a generalization can be made over ISO 50001 previously and how the concepts of energy efficiency and management can be developed. Based on the actual data of a ship, an energy efficiency analysis was conducted, and the performance evaluation was discussed. The improvement potentials were also evaluated by evaluating the travel processes of the ship and the consequent consumption. The holistic potential of the ship was evaluated by examining the possible effects on the improvement potential along with the consumption-based emission distribution of the ship. At the end of the study, the ship's energy management preferences and energy efficiency potentials were evaluated based on the analysis.

10.2 Maritime Sector and Energy Projection

The maritime industry has a structure that is based on awareness in the fight against climate change in a global sense and shapes the processes with rules. As a matter of fact, the International Maritime Organization (IMO) defined the energy demand of the industry globally as approximately 11 exajoules (EJ) in its Fourth Greenhouse Gas Study conducted in 2020. This potential represents approximately 3% of the total potential with 1 billion tons of carbon dioxide (CO_2). Although there is a variety of fuels in the maritime sector, the predominant fuel type is fossil fuels. This represents 99% of total consumption, using heavy fuel oil (HFO), marine gas oil (MGO), very low sulfur fuel oil (VLSFO), and liquefied natural gas (LNG). While the maritime sector is responsible for approximately 90% of the total trade in international shipping, it also corresponds to approximately 70% of the total energy-related emissions in emission generation (Bertzeletou 2021). Development scenarios expect an annual increase of 4% in commodities in maritime transport, based on 2050. This approach means a threefold increase in the emission potential under current conditions (James and James 2008). In this context, IMO has developed pioneering targets for 2050. Particularly, in maritime transport, it bases the decarbonization target on a quest toward net zero as the main action. This perspective, which is based on reducing carbon intensity, basically defines a search in two directions, operational, and technical. For this purpose, for the minimum carbon intensity target defined by IMO 2030, ships have been discussed from two perspectives. From an operational and technical point of view on engines, it is valuable to limit the main engine power and improve energy efficiency in ski ships.

The use of fossil fuels in maritime transportation has been the subject of extensive research in terms of environmental threat, decarbonization, and sustainability. IMO has developed important regulations on this problem with carbon management models, including sulfur control (Tadros et al. 2019). For this purpose, energy efficiency and effective management of energy are a priority approach. Energy efficient management, which will be developed as a management tool on ships, has made dynamic planning and implementation processes mandatory in order to ensure the sustainability of energy efficiency.

10.3 Energy Management for Ships

Ships, as living organizations, have dynamic and complex structural features with different operational processes. Ships have very different management standards within an international organizational structure. Energy, which is the most important input for ships, is the priority issue that all management elements take into account in these processes. It is used for very different needs, such as energy consumption, heat, electricity, and power on ships, which basically have a system structure connected to a single source. This structure needs an effective and dynamic structure to ensure its effectiveness as a sustainable structure. The SEEMP plan, which is used as an IMO requirement, is a requirement applied to almost all ships for this purpose and is a management plan sought in processes.

Energy management on ships, in short, is a service flow based on collecting consumption data of original energy users, finding efficiency potential by considering the efficiency points of the data, finding problem areas that cause consumption inefficiency, preparing an action plan for this purpose, and monitoring and monitoring processes based on continuity. In this context, SEEMP does not contain very effective features along with some areas of responsibility. Today, it is seen that three management models, namely the ISM code, SEEMP, and ISO50001, are applied in corporate practices. However, there are differences in terms of basic criteria. In this context, Fig. 10.1 offers a comparison.

Despite different types of applications, ISO 50001 energy management system has found many sectoral areas, while EEDI and SEEMP have been applied in the maritime sector. However, the most important process is the development of a management strategy that will guide them in ship management (Armstrong and Banks 2015). International Convention for the Safety of Life at Sea (SOLAS), International Convention

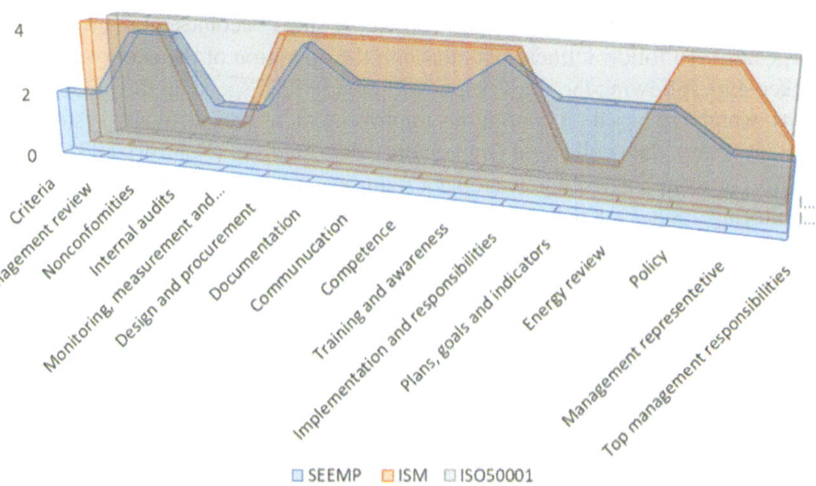

Fig. 10.1 Comparative of energy management systems

for the Prevention of Pollution from Ships (MARPOL), and International Convention on Standards of Training, Certification, and Watch keeping for Seafarers (STCW) regulations have been developed to manage IMO international processes. Among them, MARPOL is the most important. It is basically an international convention for the prevention of pollution on ships (Campara et al. 2018). However, IMO has introduced many criteria for energy efficiency institutionally. Developed under the Marine Environment Protection Committee (MEPC), the Energy Efficiency Design Index (EEDI), Energy Efficiency Operational Index (EEOI), Energy Efficiency Existing Ship Index (EEXI), and Carbon Intensity Indicator (CII) criteria have been developed for energy efficiency and carbon monitoring on ships. However, as the basic criteria in energy management, efficiency measures related to power management should also be taken into account.

10.4 Methodology

In this study, the energy management system was modeled with reference to a tanker ship, and a framework based on the evaluation of energy efficiency was developed. As a process analysis of this framework, an approach has been developed directly over the ISO 50001 energy management standard. The effects on thermal performance, primarily on energy efficiency, were evaluated. In addition, sectoral performance evaluations over EEDI and EEOI criteria were also taken into account in the study. The applicability of ISO 50001 varies according to the organizational structure, objectives, informational resources, and technical resources within the energy management system. This does not prevent the creation of every energy management system, but it does set different requirements for almost every PDCA cycle step. Thanks to the PDCA cycle provided by ISO 50001, the development of Energy Management System (EnMS) on tanker ships becomes more organized and purposeful. To follow a linear path through the operation of the energy system of a tanker ship, following the steps below will greatly help speed up the process. These steps, which are based on continuous improvement, are respectively; system development (Plan), data collection (Implement), data analysis (Control), and action and results (Activate). Energy management requires a proper monitoring phase, having monitoring tools that help to examine and analyze the efficiency of the energy management system. Monitoring tools may differ or be changed according to the plan. Efficiency measures on ships can be carried out with different applications and criteria. However, the IMO criteria are given below. According to EEDI,

$$EEDI = \frac{W.SFC.F_{CO2}}{C_{Dwt}.V_{ship}} \tag{10.1}$$

where W is power of the ship, SFC is specific fuel consumption, C_{Dwt} is capacity of the ship, and V_{ship} is velocity of the ship. Indicators recommended by IMO, such as the EEOI, could be the main energy efficiency measures. The ship's EEDI can be a reference point for the EEOI during voyages. In addition, the state of SEU equipment such as the main engine can be measured by thermal efficiency calculation. Indicators recommended by IMO, such as EEOI, can be the main energy efficiency measures of ships. EEOI is an operational indicator that was developed by MEPC and helps the ship-owners, operators, and parties concerned in the evaluation of energy efficiency regarding CO_2 emissions. Ships' emissions directly come from the fuel used in machinery (IMO 2009). Accordingly, EEOI;

$$EEOI = \frac{\sum_j FC_j x C_{f\,j}}{m_{cargo} x D}, \quad or \ Average \ EEOI = \frac{\sum_i \sum_j FC_j x C_{f\,j}}{\sum_i m_{cargo} x D} \tag{10.2}$$

where FC is fuel consumption, CF is fuel-related CO_2 emission factor, m is cargo load, and D is distance in miles, indices of i and J state to fuel and voyage, respectively.

In thermodynamics, the thermal efficiency is a performance indicator for machines that use thermal energy, such as internal combustion engine for ships. Simply, thermal efficiency is the ratio between the useful output and the input of a process that shows the losses from the input. As far as we know from the first and second laws of thermodynamics, the energy output cannot be equal or more than the input in non-ideal processes.

$$\eta = \frac{\dot{W}_{net}}{\dot{Q}_{in}} \tag{10.3}$$

where η_{th} is the thermal efficiency, W_{net} is the output work of the engine, and Q_{in} is the input energy coming from the fuel.

10.5 Results and Discussion

In this study, the energy efficiency and ISO50001-based energy management framework for ships, which have a deadweight of 34,994 metric tons, are examined. It was evaluated with reference to a chemical tanker with a length of approximately 186 m. Energy efficiency is seen as an important criterion in terms of tankers operating in ECA waters, navigation areas, and environmental pollution caused by emissions. It is the operation of machinery in accordance with the ship's audited management guidelines and regulations, ISM, SEEMP, or ISO50001, to achieve energy efficiency and proper energy management. In this context, a framework has been created for ISO50001. In this part of the study, how ISO 50001 can be integrated on a tanker ship according to ISO 50001 requirements is examined. In this context, an energy management plan was first presented for the tanker (Table 10.1).

The fundamental step in energy management is the identification of energy efficiency potential. In this context, the fuel performance of the reference tanker ship was evaluated, and it was seen that the basic indicator was the distance. Based on

Table 10.1 Energy management onboard based on ISO50001

Flow steps	Scope	Responsibilities
Management Responsibility	An institutional requirement and the appointment of a manager	Management
EnMS Manager	Master; Person responsible for the effective implementation of EnMS on board	Management officer
Energy Management Team	Captain/Captains, Chief engineer and section engineers	Management officer
Energy Policy	Energy manager and management according to ISO 50001 with EnMs sorumlusu tarafindan yazilacak ve duyurulacak	EnMS Management Onboard and management officer
Energy Review	Specific Energy Users (SEUs) do what is done according to the plans of the Goals	EnMS Manager Onboard
Responsibilities	Administrative sub-management servants related to all operations from the captain	EnMS Manager Onboard
Education and Awareness	All trainings according to ISO 50001 Gemini regulation planning and implementation	EnMS Manager Onboard
Competence	Energy competence parameters; Psychometrics, Simulation, Group work, Briefing exercise	EnMS Manager Onboard
Communication	Communication between the ship owner or the Customer, Company, partners, Captain and ship personnel regarding processes and workflows	EnMS Manager Onboard
Documents	Necessary documents related to IS050001 and SEEMP; MEPC.1/Circ.683, MEPC Guidelines for voluntary use of EEOI, MARPOL annex Vi and other issuse	EnMS Manager Onboard
Operational control	Ship operations related to SEU's operational parameters and ship energy efficiency	Process management
Monitoring, Measurement and Analysis	Monitoring processes and procedures for SEU and ship operations	EnMS Manager Onboard and process management

(continued)

Table 10.1 (continued)

Flow steps	Scope	Responsibilities
Management review	All EnMS process on board and improvement opportunities comprehensive review based on	EnMs management onboard and management officer

this, 48 travel times were taken as a basis, and the energy efficiency potential was investigated with regression analyses, and the results are given in Fig. 10.2.

When examined over the target consumption equation related to the ship's travel processes, the energy efficiency potential was found to be 10.17%. The average load of the main engine of the ship operating at low performance load was found to be 73.12%. The main indicators to be monitored for energy management are examined for the reference ship, and the distributions are presented in Fig. 10.3.

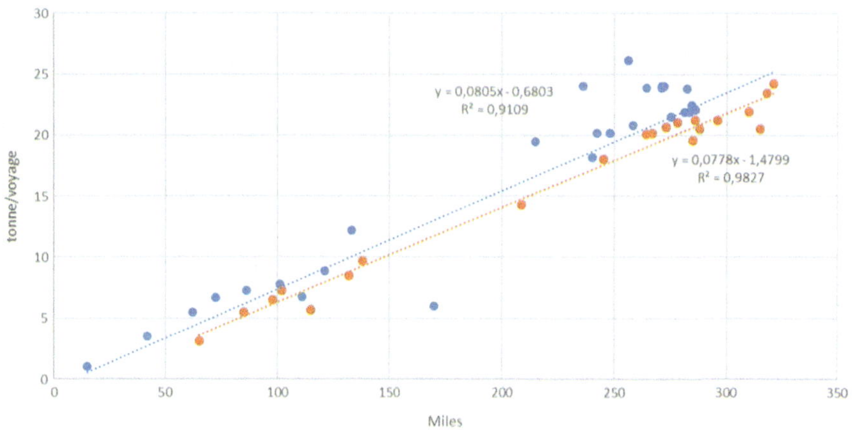

Fig. 10.2 Energy consumption distributions of ship with regression analysis

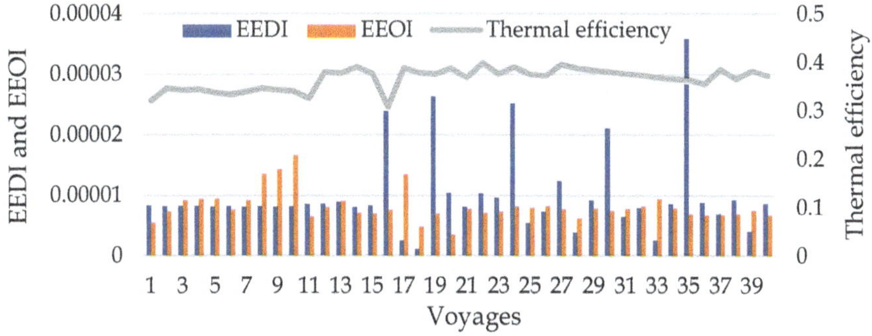

Fig. 10.3 Energy performance indicators for the tanker ship

Performance criteria differ within themselves. Considering the averages, EEDI has a variation fluctuation between 3% and −250%, while this value varies between − = 10% and 56% in EEOI. However, this fluctuation in thermal efficiency indicates a value of approximately ± 10. In this context, it can be said that the thermal efficiency has a more effective range in order to monitor the efficiency of the efficiency.

10.6 Conclusion

This study basically presents a developed framework to effectively examine energy management and energy efficiency in ships. In this context, an approach evaluation was made for the development of the infrastructure regarding the applicability of ISO 50001 on ships and for energy efficiency. For this purpose,

- The energy efficiency potential for 48 trips was found to be approximately 10.17. This can be defined as the target for energy management.
- The EEDI average for the ship was found to be 1.0169E—05 (g CO_2/ton.mile (cargo carried)), while the EEOI average was found as 8.3402e-06 (tons CO_2/(tons nautical miles)).
- Thermal efficiency is important in terms of generating unit power. As a matter of fact, while the general performance average is 36.48%, the value fluctuation shows a change of ±10. This is a more controlled traceable value.

The maritime industry can be considered as an effective model for more effective and corporate solutions with the ISO 50001 Energy Management System within the traditional structure. This type of work can be supported by operational process management and entropy management for this purpose.

References

Adland RO, Cariou P, Jia H, Wolff F (2018a) The energy efficiency effects of periodic ship hull cleaning. J Clean Prod 178:1–13

Adland R, Cariou P, Jia H, Wolff F-C (2018b) The energy efficiency effects of periodic ship hull cleaning. J Clean Prod 178(2018):1–13

Armstrong Victor N, Banks Charlotte (2015) Integrated approach to vessel energy efficiency. Ocean Eng 110, Part B, pp 39–48, ISSN: 0029-8018, https://doi.org/10.1016/j.oceaneng.2015.10.024

Baldi F, Johnson H, Gabrielii CH, Andersson K (2014) Energy and exergy analysis of ship energy systems—the case study of a chemical tanker

Bertzeletou M (2021) On the path to decarbonize shipping and Reducing CO_2 emissions, The Signal Ocean platform, London, UK

Čampara L, Hasanspahić N, Vujičić S (2018) Overview of MARPOL ANNEX VI regulations for prevention ofair pollution from marine diesel engines. SHS Web of Conf 58(5–6):01004. https://doi.org/10.1051/shsconf/20185801004

Georgopoulou C, Koukoulopoulos L, Dimopoulos G (2021) Model-based systems engineering for the design and operational assessment of marine energy systems and retrofitting solutions. A holistic approach to ship design

https://www.thesignalgroup.com/newsroom/maritime-decarbonization-reducing-co2-emissions-in-shipping

https://www.classnk.or.jp/hp/pdf/activities/statutory/eedi/mepc_1-circ_684.pdf

https://www.imo.org/en/MediaCentre/PressBriefings/Pages/06GHGinitialstrategy.aspx

https://www.classnk.or.jp/hp/pdf/activities/statutory/seemp/seemp-mepc282-70.pdf

IEA, CO_2 emissions from fuel combustion highlights, 2019, International Energy Agency OECD: Paris, France, https://iea.blob.core.windows.net/assets/eb3b2e8d-28e0-47fd-a8ba-160f7ed42bc3/CO2_Emissions_from_Fuel_Combustion_2019_Highlights.pdf

IMO (2009) Guidelines for voluntary use of the ship energy efficiency operational indicator (EEOI), MEPC.1/Circ.684, London, UK

IMO (2018) UN body adopts climate change strategy for shipping. International Maritime Organisation, London

James JC, James W (2008) The impacts of globalization on international maritime transport activity past trends and future perspectives. Energy environmental research associates, the United States

Kalli J, Karvonen T, Makkonen T (2009) Sulphur content in ships bunker fuel in 2015: a study on the impacts of the new IMO regulations and transportation costs. (Publications of the Ministry of Transport and Communications; Vol No. 31)

Knorring Von H, Johansson M, Andersson K, Södahl B (2012) Will the IMO ship energy efficiency management plan (SEEMP) lead to reduced CO_2 emissions? A comparison with ISO 50001 and the ISM Code

MEPC (2016) Guidelines for the development of a ship energy efficiency management plan (seemp), annex 10, resolution MEPC 282(70)

Perera LP, Mo B (2016) Emission control based energy efficiency measures in ship operations. Appl Ocean Res 60:29–46

Rehmatulla N, Smith TW (2015) Barriers to energy efficiency in shipping: a triangulated approach to investigate the principal agent problem. Energy Policy 84:44–57

Seddiek IS, Elgohary MM (2014) Eco-friendly selection of ship emissions reduction strategies with emphasis on SOx and NOx emissions. Int J Naval Archit Ocean Eng 6(3):737–748, ISSN: 2092-6782

Tadros M, Ventura M, Guedes Soares C (2019) Optimization procedure to minimizefuel consumption of a four-stroke marine turbocharged diesel engine. Energy 168(2019):897–908

Chapter 11
Evaluation of the Market-Based Measures by the IMO Criteria: Effects of Current Dynamics

Ufuk Yakup Çalışkan and Burak Zincir

Abstract Market-based Measures (MBM) are deemed one of the feasible methods to achieve sustainable maritime transportation in the mid-term by International Maritime Organization (IMO). Recently, shelved discussions of MBMs reopened at the Marine Environment Protection Committee's 76th meeting to possibly apply to shipping operators after European Commission decided on including shipping in Emission Trading Scheme (ETS). This chapter will go over the proposed candidate MBMs and address the possible drawbacks of each emission reduction scheme. Included are background information, a comparison of MBMs, an illustration of the impact of carbon pricing and fuel levies on shipping operators, and policy improvements. Assessment made on their effectiveness considering their harmony to the existing legal framework, availability of the implementation in terms of time windows, impact on various states in terms of development and geographical disadvantages, administrative burden, practical feasibility, and impact on the profitability. Outcomes interpreted over today's conditions are listed. Within these possibilities, the best MBM scenarios have been tried to be drawn. As a result of the interpretation, medium-level levy and low-to-medium-level ETS were the most reasonable options based on the literature. Levy and ETS are the most important among MBMs. In addition to this, levy still has a significant advantage over ETS.

Keywords Market-based measures · Carbon tax · Bunker levy · Emission trading scheme · Carbon pricing

U. Y. Çalışkan (✉)
Division of Transportation Services. Liman, Bartin University, Iskele Cad. Kurucasile, 74500 Bartin, Turkey
e-mail: ucaliskan@bartin.edu.tr

B. Zincir
Maritime Faculty, Postane Mah. Sahil Cad. Tuzla, Istanbul Technical University, 34940 Istanbul, Turkey
e-mail: bzincir@itu.edu.tr

B. Zincir et al. (eds.), *Decarbonization of Maritime Transport*, Energy, Environment, and Sustainability, https://doi.org/10.1007/978-981-99-1677-1_11

Abbreviations

CII	Carbon Intensity Indicator
CH_4	Methane
CO_2	Carbon dioxide
CO_2e	CO_2 equivalent
COP	Conference of Parties
DCS	Data Collection System
EEDI	Energy Efficiency Design Index
EEXI	Energy Efficiency Existing Ship Index
EIS	Efficiency Incentive Scheme
EU	European Union
ETS	Emission Trading Scheme
FOE	Friends of the Earth
GHG	Greenhouse Gas
ICS	International Chamber of Shipping
IMF	International Monetary Fund
IMO	International Maritime Organization
ISWG	Intersessional Working Group
ITF	International Transport Forum
IUCN	International Union for the Conservation of Nature
LDC	Least developed countries
LIS	Leveraged Incentive Scheme
LNG	Liquefied Natural Gas
MACC	Marginal Abatement Cost Curve
MBM	Market-Based Measures
MGO	Marine Gas Oil
MPEC	Marine Environment Protection Committee
MRV	Monitoring Reporting and Verification
NOx	Nitrogen Oxides
OECD	Organization for Economic Co-operation and Development
PM	Particulate Matters
PSL	Port State Levy
RM	Rebate Mechanism
SECT	Ship Efficiency and Credit Trading
SIDS	Small Island Development States
SOx	Sulfur Oxides
TRL	Technology Readiness Level
UN	United Nations
UNFCCC	United Nations Framework Convention on Climate Change
US CBO	United States Congressional Budget Office
USD	United States Dollars
VES	Vessel Efficiency System
WSC	World Shipping Council

11.1 Introduction

According to the International Maritime Organization's (IMO) 4th Greenhouse Gas (GHG) emissions inventory study (IMO 2018), share of shipping in the global anthropogenic emissions has increased from 2.76% in 2012 to 2.89% in 2018. This includes the unexpected increase in methane (CH_4) emissions by 2.5 times and the expected one for carbon dioxide (CO_2) emissions by 5.6%. More concerningly, emissions of the air pollutants inside of ship plumes such as nitrogen oxides (NOx), sulfur oxides (SOx), and fine Particulate Matters (PMs) increased as well in the same time window by 1.2%, 5.5%, and 3.6%, respectively. The goal setting to reduce shipborne GHG emissions by IMO took place in 2018 (IMO 2021a). The strategies are categorized under three approaches: technical, operational, and market-based (Shi and Gullett 2018). In terms of Market-based Measures (MBM), many variants of schemes to promote a greener perspective among shipping operators are proposed. These include the most known levy on bunker fuel and European Union's (EU) Emission Trading Scheme (ETS) (Psaraftis et al. 2021).

In the light of recent changes such as the inclusion of shipping into EU ETS, tangible perspective change within the Marine Environment Protection Committee (MEPC), and the hikes in fuel and lube oil costs, MBM literature should be revisited by a sound set of criteria (European Commission 2021; EU 2019; Marine Link 2022; Psaraftis 2021; Kırval and Çalışkan 2022). The technical evaluations and comparisons made on the subject were either made a decade ago or conducted without being able to predict conditions would change beyond recognition. Thus, this study brings a perspective to the MBM literature with today's conditions and evaluates new suggestions and established proposals before a historical IMO decision. In this section, we first discuss the background of MBMs and their literature. In the latter, the potential effectiveness of the systematic schemes is evaluated via a literature review on the comparisons among MBMs. Then we try to shed some light on how current dynamics can reshape the MBMs and the proposals of states and important maritime actors. Policy implications are included in the last discussions.

11.2 The Background

In former times, companies that have polluted have not paid the cost of that pollution, which brought the need for an active environmental law. One of the six major principles in environmental law is the principle of "polluters pay". The institution of this principle goes back to Rio de Janeiro 1992 United Nations (UN) conference, and prior to that many policy efforts to put the principle into its final shape (Beder 2013; United Nations Conference on Environment and Development 1992). Discussions of internalization of the environmental cost of maritime transportation have been around since 1995 (Lagouvardou et al. 2020). In the 45th meeting of

MEPC, the long-lasting argumentation on potential MBM started with an assessment within the context of a GHG inventory report (Skjølsvik et al. 2000). After ten years, in the 60th meeting of MEPC, 11 proposals of MBM have been brought to the table by several member states of the IMO and IMO observer organizations. An expert group was assigned to assess the possibility of applying measures that catalyze in-sector or out-of-sector (offsetting) emission reduction. Eventually, MBM discussions were suspended after three years (Psaraftis et al. 2021). At the 72nd MEPC meeting, priorly envisioned adoption of the Initial IMO Strategy took place. A list of candidate measures is introduced to reach the Paris Agreement goals of dragging global average temperature increase to 2 °C and below in this process. The list was constructed on three time windows: short-term (2018–2023), mid-term (2023–2030), and long-term (2030–2050) (IMO 2018). Follow-up actions to Initial GHG Strategies discussed at 4th and 5th Intersessional Working Group GHG (ISWG-GHG) 4th and 5th meeting and 73rd and 74th MEPC sessions. Activities to deal with medium and long-term candidate actions with their respective identified barriers, impacts on states, and the 4th IMO GHG study were deemed attention-worthy (Shi and Gullett 2018; Joung et al. 2020; Christodoulou et al. 2021). Although MBMs are mentioned under mid-term measures, some MBM proposals can incentivize alternative marine fuels, which are considered a long-term measure, and inducing slow steaming, considered a short-term measure (Lagouvardou et al. 2020).

Until recently, IMO was not inclined toward MBMs. The solidified decision of the EU may have inspired states in the 76th MEPC meeting. Eventually, three-phased approaches to consider and develop such measures are agreed on. The first step will be the collation and consideration of the proposal, followed by the assessment and selection of the measure. The final step is developing a measure by 2023 (Psaraftis 2021; Kırval and Çalışkan 2022). The uncertainty caused by European conflicts and the skyrocketing prices of fossil fuels has stalled the options (Marine Link 2022). For instance, Liquefied Natural Gas (LNG) is one of the most affected fuels by the sudden increase in fuel prices worldwide. Marine Gas Oil (MGO), which was in the range of 550 to 650 United States Dollars (USD) last year, is in the range of 1350–1450 USD in June (1160 on 06.07.2021). The metric ton of LNG, which was 2705 USD at the port of Rotterdam, was 778 USD on the same day last year (06.07.2021–2022) (Ship & Bunker 2022). This gap of 1545 USD interrupts the policy of reducing the impact of maritime transport on air pollution by using alternative fuels. Even though LNG has high calorific values, an energy difference of 15.22% is not enough to close the price deficit of air polluting and alternative fuels (Gil-Lopez and Verdu-Vazquez 2021).

11.2.1 MBM Proposals

The nine criteria were given to the expert group to evaluate the original ten proposals. The set of criteria consisted of:

(1) environmental effectiveness,
(2) cost-effectiveness and potential impact on sustainable development and trade,
(3) potential catalyzer effect of MBM on innovation and technological improvements,
(4) practicality,
(5) the need of technology transfer and capacity building for the developing countries, especially the least developed countries (LDCs) and the Small Island Development States (SIDS),
(6) the relevancy to international and supranational environmental and other conventions such as the United Nations Framework Convention on Climate Change, Kyoto Protocol, and United Nations Convention on the Law of the Sea,
(7) administrative burden and legal aspects,
(8) additional workload, economic burden, and operational impact for individual ships, the shipping industry, and the maritime sector as a whole, and
(9) compatibility with existing policy framework of the proposed MBM (IMO 2010).

In addition to ten proposals, after canceling the discussions, several suggestions were made by states and researchers. Table 11.1 contains the list of proposals and suggestions and their shallow breakdowns.

In their core state, four of the proposals have been standing out: bunker levy, ETS, PSL, and RM as an add-on to any applied MBM. GHG Fund schemes and functionality of schemes based on the GHG Fund such as EIS and PSL can be associated with bunker levy. At the same time, ETS can be related to EU ETS (Chai et al. 2019).

11.2.2 Comparison of MBMs in the Literature

Levy is denoted as superior to ETS in the literature by being more flexible, easy to monitor, and simple (Lema et al. 2017). While PSL is unable to tackle emissions as much as it intended. Although ETS tackles emissions, the complexity that arises with the administrative burden and need for the regulatory body for the states that are not part of the EU can struggle in the such scheme (Psaraftis and Lagouvardou 2019). US Congressional Budget Office (CBO) (2008) indicated that an ETS would be twice as costlier than implementing a levy. The market operators have given the same outlook on comparing the levy and ETS with the literature in the perspective of being more effective and simplistic (Giziakis and Christodoulou

Table 11.1 MBM proposals and suggestions by their original contributor and their brief explanation (Bahamas 2010; Cyprus et al. 2010; Devanney 2010; France 2010; Germany 2010; IUCN 2010; Jamaica 2010; Japan 2010; Norway 2010; United Kingdom 2010; United States of America 2010; WSC 2010; IMO 2011; Japan and WSC 2011; Kågeson 2011; Gkonis and Psaraftis 2012; Cristea et al. 2013; Parry et al. 2018; IMO 2019; Psaraftis 2019a; Psaraftis 2019b; Tanaka and Okada 2019; Wang et al. 2019; Trafigura 2020; IMO 2021b; IMO 2021c; Maritime Magazine 2021; Psaraftis et al. 2021; Tiwari et al. 2021)

Scheme	Proposed by	Brief explanation
Market-Based Instruments	Bahamas (2010)	• Proportionately penalty on trade and development • Based on the Energy Efficiency Design Index (EEDI), including domestic maritime transportation • Withdrawn
GHG Fund	Cyprus et al. (2010)	• If the global GHG reduction goal is not met, offsetting financed by a contribution paid by ships on every ton of bunker fuel purchased should occur • Original version of bunker levy proposal
Carbon tax	Devanney (2010)	• Carbon tax applied to EEDI calculations • 155 USD
Further elements for the development of an ETS	France (2010)	• Design of the auctioning is proposed on Norway's proposal
Global Emission Trading System (ETS)	Germany (2010)	• Similar to ETS proposals
Rebate Mechanism	International Union for the Conservation of Nature (IUCN) (2010)	• Can be an add-on proposal to any MBM that calculating the loss of developing countries that such MBM can harm in terms of economic activities

(continued)

Table 11.1 (continued)

Scheme	Proposed by	Brief explanation
Port State Levy (PSL)	Jamaica (2010)	• Calculating emissions on a direct GHG emission reduction scheme from ships through port state arrangements • Levy is applied not on the bunker supplies but in the aftermath of the voyage's GHG impact in the same manner as Japan's proposal
Leveraged Incentive Scheme (LIS)	Japan (2010)	• Direct reduction of CO_2 is targeted by creating performance benchmarks and labeling ships as a good or bad performers within the context of energy efficiency • Similar to the Energy Efficiency Existing Ship Index (EEXI) and Carbon Intensity Indicator (CII) measures but based on the GHG Fund (IMO 2021b)
Global Emissions Trading System (ETS)	Norway (2010)	• Allowances called Ship Emission Units are assigned each year and will be released into the market • Shipping operators should stay within the pre-disposed target emission cap in a sector-wide scheme • Allowing operators to use out-of-sector credits beside in-sector credits, which enables vertical integration with other nodes in maritime transportation or greater supply chain • Auctioning revenue generated should be assigned to the development of cleaner technology and goals of adaptation and mitigation • Limited exceptions may be applicable for SIDS • Applicable to fossil-fuels propelled ships that carry out international trade above certain cargo volume • Along the lines of EU ETS

(continued)

Table 11.1 (continued)

Scheme	Proposed by	Brief explanation
Global Emissions Trading System	United Kingdom (2010)	• The approach to setting emission caps is different than Norway's proposal • Generated auctioning revenues will remain with national governments
Ship Efficiency and Credit Trading (SECT)	United States (2010)	• Rather than applying a surcharge on bunker fuel or a cap-and-trade scheme, a series of reduction activities should be applied via an efficiency-credit trading program on an established baseline • Similar to the phased approach of the Energy Efficiency Design Index (EEDI), the reduction plans get stricter in each phase (IMO 2011)
Vessel Efficiency System (VES)	World Shipping Council (WSC) (2010)	• Similar to EEDI consisting of levels of standardized efficiency getting stringent over time • Non-compliant ships will be subject to pay a fee
Efficiency Incentive Scheme (EIS)	Japan and WSC (2011)	• Combination of LIS and VES proposals
Carbon tax	Kågeson (2011)	• Gradual increase until 2025 • Starting from the base level of 30 USD • Around 50 USD
Bunker Levy	Gkonis and Psaraftis (2012)	• 425 USD to make a ton of bunker price 1,000 USD

(continued)

Table 11.1 (continued)

Scheme	Proposed by	Brief explanation
Carbon tax	Cristea et al. (2013)	• Approximately 160,3 USD per ton of bunker fuel
Carbon tax	Parry et al. for International Monetary Fund (IMF) (2018)	• In the form of a carbon tax of 5 to 30 USD
Bunker Levy	International Chamber of Shipping (ICS) et al. (IMO 2019)	• 2 USD worth of levy on bunker fuel
Bunker Levy	Pacific Islands (Psaraftis 2019a)	• 2 USD worth of levy on bunker fuel
Bunker Levy	Psaraftis (2019b)	• 425 USD to make a ton of bunker price 1,000 USD
Carbon tax	Tanaka and Okada (2019)	• An incentivizing CO_2 tax scheme to make shipping business adopt a lesser polluting path
Bunker Levy	Wang et al. (2019)	• To make ton of bunker price double from 300 to 600 USD

(continued)

Table 11.1 (continued)

Scheme	Proposed by	Brief explanation
Carbon tax	Trafigura (2020)	• 801,5 to 961,8 USD tax per ton of bunker fuel • As the highest levy in literature per ton of CO_2
Carbon tax	Marshall Islands and Solomon Islands (IMO 2021c)	• 320.6 USD or 100 USD CO_2 tax
Bunker Levy	Maersk (Maritime Magazine 2021)	• 450 USD to make a ton of fossil fuel prices exceed the prices of green fuels
Carbon tax	Tiwari et al. (2021)	• 96,18 USD • Optimal version among business-as-usual, 10, 30, and 50 USD carbon tax scenarios
A short list of GHG Fund and ETS proposals	Greece (Psaraftis et al. 2021)	• Turned down

2012). In Table 11.2, the studies and proposals are categorized on the basis of either proposing the respective MBM, the result of their calculations, finding effective positive outcomes of respective MBM, or depending on their suggestion of choice among MBMs.

11.3 Methodology

This paper adopts an updated perspective with recent changes to cover the original nine criteria. Instead of the long criterion names, short representative criterion names are assigned. The comparison is mainly deduced in favor of either the ETS or bunker levy since the main diversion is drawn between them.

11.3.1 The Nine Criteria

As the first step, MBMs are revisited based on the nine criteria in the current dynamics of maritime business. The criteria order is as in the previous work (IMO 2010). Each criterion is evaluated with its relation or intersecting points to other criteria. Most criteria included the comparison of ETS and bunker levy since these measures are held most probable by all actors.

Environmental Effectiveness. If the ecological return and honoring all the objectives are aimed, the levy should be set in a higher position (Kågeson 2011). Although medium to high-level levies lead to slow steaming and various offsetting trade movements, only robust levy levels can pay off in abandoning fossil fuels (Wang et al. 2019; Kachi et al. 2019). On the other hand, ETS primarily focused on reducing emissions (Faber et al. 2009). Complete certainty on the cap will eventually give results as it has been (Psaraftis 2012). A direct subsidy may be the remedy for the CO_2 tax via funding efforts to upgrade fuel efficiency (Tanaka and Okada 2019).

From the environmental perspective, even though maritime transportation is believed to be highly regulated, the urgent need for an MBM surpasses the search requirement for the best suitable action (Faber et al. 2009). The business-as-usual state is projected to increase CO_2 emissions to half of what is currently emitted (Carlo et al. 2020). According to Intergovernmental Panel on Climate Change's 6th assessment report (Intergovernmental Panel on Climate Change 2021), in the last decade (2011–2020), temperatures exceeded those of the most recent multi-century warm period, around 6.500 years ago, reaching the smallest Arctic Sea ice area since at least the past 1.000 years. When the action on both MBM's grounds is believed to be helpful, the dwelling may cause irreversible damage.

Cost-Effectiveness. Lema et al.'s (2017) comparison indicates the most effective MBMs among all are ETS and a levy in tackling emissions; however, owing to its elementariness and semi-self-sustaining nature, levy variants will hold the advantage

Table 11.2 Literature on MBM proposals by favoring (CBO 2008; Faber et al. 2009; FOE 2009; High-level Advisory Group on Climate Change Financing 2009; Bahamas 2010; Cyprus et al. 2010; Devanney 2010; Ellerman et al. 2010; France 2010; FOE 2010; Germany 2010; IUCN 2010; Jamaica 2010; Japan. 2010; Norway 2010; United Kingdom 2010; United States 2010; WSC 2010; Japan and WSC 2011; Kågeson 2011; Miola et al. 2011; Cariou and Cheaitou 2012; Giziakis and Christodoulou 2012; Gkonis and Psaraftis 2012; Psaraftis 2012; Cristea et al. 2013; Kapetanis et al. 2014; Wang and Xu 2015; Shi 2016; Kosmas and Acciaro 2017; Lema et al. 2017; Wang and Chen 2017; Avetisyan 2018; ICS 2018; ITF/OECD 2018; Parry et al. 2018; BHP Group 2019; Chai et al. 2019; Gu et al. 2019; Halim et al. 2019; IMO 2019; Kachi et al. 2019; Psaraftis 2019a; Psaraftis 2019b; Psaraftis and Lagouvardou 2019; Tanaka and Okada 2019; Trivyza et al. 2019; Wang et al. 2019; Carlo et al. 2020; Tillig et al. 2020; Trafigura 2020; Christodoulou et al. 2021; IMO 2021c; Maritime Magazine 2021; Mundaca et al. 2021; Psaraftis et al. 2021; Tiwari et al. 2021; Parry et al. 2022)

Bunker levy	ETS	Other studies
CBO (2008)	Faber et al. (2009)	Bahamas (2010)
FOE (2009)	Ellerman et al. (2010)	Devanney (2010)
High-level Advisory Group on Climate Change Financing (2009)	France (2010)	International Union for the Conservation of Nature IUCN (2010)
Cyprus et al. (2010)	Germany (2010)	United States (2010)
FOE (2010)	Norway (2010)	Cariou and Cheaitou (2012)
Jamaica (2010)	United Kingdom (2010)	Psaraftis (2017b)
Japan (2010)	Kågeson (2011)	Wang and Xu (2015)
World Shipping Council (WSC 2010)	Miola et al. (2011)	Wang and Chen (2017)
Japan and WSC (2011)	Lema et al. (2017)	Avestiyan (2018)
Giziakis and Christodoulou (2012)		Gu et al. (2019)
Gkonis and Psaraftis (2012)		Halim et al. (2019)
Cristea et al. (2013)		Tanaka and Okada (2019)
Kapetanis et al. (2014)		Trivyza et al. (2019)
Shi (2016)		Carlo et al. (2020)
Kosmas and Acciaro (2017)		Tillig et al. (2020)
International Chamber of Shipping (ICS) (2018)		
International Transport Forum (ITF) and Organisation for Economic Co-operation and Development (OECD) (2018)		
Parry et al. for International Monetary Fund (IMF 2018)		
BHP Group et al. (2019)		
Chai et al. (2019)		
ICS et al. (IMO 2019)		
Kachi et al. for New Climate Institute (2019)		

(continued)

Table 11.2 (continued)

Bunker levy	ETS	Other studies
Pacific Islands (Psaraftis 2019a)		
Psaraftis (2019b)		
Psaraftis and Lagouvardou (2019)		
Wang et al. (2019)		
Trafigura (2020)		
Christodoulou et al. (2021)		
Marshall Islands and Solomon Islands (IMO 2021c)		
Maersk (Maritime Magazine 2021)		
Mundaca et al. (2021)		
Psaraftis et al. (2021		
Tiwari et al. (2021)		
Parry et al. (2022)		

(Psaraftis and Lagouvardou 2019). Compared to ETS, a levy is considered two times more cost-efficient (CBO study on cap and trade (Issue February). 2008). On the pro-ETS side, there is a proven success (Miola et al. 2011; Ellerman et al. 2010; Wang and Chen 2017).

The fluctuation in the bunker prices, such as nowadays, can critically harm ETS, whereas the elasticity of the levy would be more allowing. Global economic bottlenecks where inflation rises almost in any country can stall progress. In short, while the ETS is a candidate to get results by serving the target in the context of cost-effectiveness, the levy promises to provide this at a lower cost but in a riskier position.

Most of the Marginal Abatement Cost Curves (MACCs) that have been used in the studies of cost-effectiveness can be simplified with Eqs. (11.1), (11.2), (11.3), and (11.4).

$$MAC(MBM) = \Delta NCOST(MBM)/\Delta CO_2(MBM) \tag{11.1}$$

$$\Delta NCOST(MBM) = \Delta GCOST(MBM) - \Delta FUEL(MBM) \times PFUEL \tag{11.2}$$

$$\Delta CO_2(MBM) = \Delta FUEL(MBM) \times F \tag{11.3}$$

$$MAC(MBM) = \Delta GCOST(MBM)/\Delta CO_2(MBM) - PFUEL/F \tag{11.4}$$

where MBM is the relative abatement measure, $\Delta NCOST$ is the cost deficit from implementing the measure, and ΔCO_2 is the difference between averted CO_2 from the industry and business-as-usual environmental case. $\Delta GCOST$ is the gross cost differential except for the fuel price, $\Delta FUEL$ is avoidance of using less respective fuel, $PFUEL$ is the respective fuel price, and F is the respective coefficient CO_2 emission factor. The curve yields positive outputs when Eq. (11.4) gives negative results (Psaraftis 2022). Since most fuel costs changed beyond recognition, MACC calculations are derailed. The drastic changes in fuel economy reshaped the level of the carbon tax. On the implementation of innovative technologies side, 64.08% of total CO_2 reduction was going to be contributed by using alternative fuels, followed by speed reduction by 7.54% (IMO 2021a). The question here is which MBM will be the catalyzer for alternative fuels and speed reduction.

Catalyzer Effect. Chai et al. (2019) state that shipping firms will react proactively to a known rise in fuel costs if a levy is applied. They favored the levy regardless of its collection method, based on incentivizing technology adoption with high Technology Readiness Level (TRL), low TRL technologies investment, and generating funds. In the same approach, FOE stated the inefficiency of ETS due to the incapability of being a driver of emissions and technological innovations. The risk of sub-prime carbon and over-reliance on offsetting are also worrisome (FOE 2009, 2010). Depending on these reasonings and environmental conventions stated under "Relevancy to conventions" headings that ease the offsetting process, ETS might perform less effectively than a levy in removing the unwanted impact of GHGs. However, a proper analysis of the determination of offsetting level would be more revealing.

Practicality. Practical options for MBMs are adding up to an extensive list, including GHG Fund, pure levy on bunker fuel, CO_2e (CO_2 equivalent) taxes, taxation of low energy efficiency performers, and fuel tariffs on a regional basis (Lagouvardou et al. 2020). From a fundamental perspective, the levy is transparent and easy to monitor. The scheme does not excessively rely on administrative bodies. Additionally, it lets the market choose the abatement method from day one. These are undeniable practical aspects against a relatively complicated and inflexible scheme that relies on an administrative body and capacity-building (BHP Group et al. 2019). Stagnant market conditions may block the efficient production of GHG fund, but eventually, the alternative bear the same risk.

The practicality issue lies in the infrastructure, which is built to some extent for the ETS (IMO 2021a; European Commission 2021; Kırval and Çalışkan 2022). Although a levy does not rely heavily on such infrastructures, the gap in between closed nowadays with the remaining few critical points.

Capacity Building. Although fair treatment cannot be achieved under both schemes, a policy that is the same for each partying state could be more likely to be agreed upon (Cristea et al. 2013). The most crucial infrastructure gaps between MBMs can be summarized as

(1) the imbalance in the capacity to carry the tax burden for LDCs,
(2) the voyage evasion for SIDS,

(3) the carbon leakage to other transport modes, and
(4) monitoring and administrative facilities.

Tax levels might hurt LDCs that carry out intercontinental transport of vegetables and fruits, where international shipping becomes costlier with increased fuel costs than intracontinental transportation (Psaraftis 2019a). Voyage evasion is when companies may reduce their voyages after an ETS decision considering the extended outcome of CO_2 emissions and relative tax to it. Carbon leakage is the modal shift to more polluted ways of transportation. Facilities refer to the bureaucratic, technologic, personnel, and scientific needs of an implied MBM.

World Bank and Ecofys (2018) noted the stark carbon pricing deficit between LDCs and developed countries. Voyage evasion can be denoted as the technical and logical flaw of ETS that should be compensated somehow for SIDS, or policy should be altered to favor SIDS and LDC (Psaraftis 2012). Within EU borders, voyage evasion and carbon leakage issues are irrelevant, but this may cause modal shifts in landlocked countries where multimodal transportation is carried out that increase in fuel cost may alter the route.

Recently, monitoring equipment developed by the IMO has facilitated the possibility of implementing the ETS system. In addition, the increase in fuel consumption reporting from ships every year is removing the need for capacity building which constitutes a significant ground for 2023. The question mark on the issue is the difference between tools. Unlike IMO's tool, the EU's tool includes time spent at sea to distance traveled, which can be a weak point against calculations of emitted carbon (DNV GL 2021).

Relevancy to Conventions. Unlike ships, production plants are not known mainly by their names but by the processes they carry out. They do not change flags. Their cultural diversity is limited. The laws they are supposed to abide are national. The maritime market is global to the point that it can hardly be compared with other markets. Therefore, their policy framework has always been designed to be universal. As a result, a unified policy framework is already in place. The problem here is the relevancy that governs trade as a whole and environmental treaties that consider all transport modes the same.

A rebate mechanism built on the GHG Fund is considered acceptable by many countries (Shi 2016). If offsetting via technical and operational methods could not be ensured, carbon leakage to different modes is likely result (ITF, OECD. 2018; Halim et al. 2019; Carlo et al. 2020). However, a unified global tax on all transport modes can drive goods such as microchips, seeds from air transport, paddy rice, and wheat and cereal grains from road transport, indicating the opportunities of a global ETS (Halim et al. 2019).

Each Conference Of Parties (COP) meeting of the United Nations Framework Convention on Climate Change (UNFCCC) after the Paris agreement, the carbon market gains more strength with new set of rules. The recent COP26 brought a disputably flawed new scheme into the carbon market (Paterson 2021). As much as the offsetting capabilities of markets are increased, ETS is far from unviable. In addition, climate finance discussions foresee a future in which developed countries help cut GHG emissions by helping LDCs and SIDS.

Administrative Burden. The main difference in administrative burden between a levy and ETS is the number of people the system needs to deal with. Some maritime business actors form companies for each individual ship in their fleet. The reasoning behind this act is the limitation of liability risks. Therefore, an ETS would require dealing with far more units than a levy, which would likely deal with bunkering businesses and fewer actors (High-level Advisory Group 2009; ITF, OECD. 2018).

The fact that the EU will not incur an extra initial investment cost with the existing ETS management is not valid for the world, as it incurs additional costs instead of abatement. Therefore, even though the ETS proposal has come a long way, the administrative advantage is still with the bunker levy.

Operational Impact. A wide range of applications among ships and certainty over prices make levy more eligible than ETS (Parry et al. 2018). Shipping companies were not enthusiastic about a high-level tax idea; however, their approach to reducing emissions is on a levy over ETS (Giziakis and Christodoulou 2012). In a volatile condition, ETS has been seen as a possible cause of market distortion (Christodoulou et al. 2021; ICS 2018). Any ETS variants arise the discussion of allowances and the level of the bar, whereas levy would be more adjustable in a timely manner (Kırval and Çalışkan 2022).

The issue of fraud is where the successful cap-and-trade system is advantageous. The obscurity of Levy's fraud prevention systems makes the ETS system, considered reliable except for a few examples, less dangerous.

Both systems are highly vulnerable to energy market disruptions. Today's energy market decisions are mainly in the shadow of political choices. Any future political disagreement may prevent both MMB decisions from being implemented. In addition, energy exporting countries may not accept a system with an administrative body for political reasons. On the other hand, since the levy's design directly deals with bunkering infrastructure, these exporting states may disrupt the scheme in a political dispute.

Regarding operational impact, ETS deserves more scrutiny with its complicated nature and policy implications arising from the issues stated under the "Capacity building" heading. The straightforward nature of the levy is another upside factor to favor the levy with certainty over prices and being applicable for all existing and new ships.

Compatibility with Existing Policy Framework. As it has always been, the nature of shipping policies demands adaptability over superiority. As the EU approved the inclusion of shipping into EU ETS, significant cargo volume that comes, goes, and moves within will be subject to an MBM. Applied rules of trade for shipping will be norms for many transportation actors. While it is beneficial for the EU to reduce emissions from shipping through the ETS, it would be wrong to assume that the same input will yield the same output as the IMO does not carry the same infrastructure. However, since the same methods or similar ones are applied at the supranational and international level, a certain harmony has been achieved, i.e., Monitoring Reporting and Verification (MRV) and IMO's Data Collection System (DCS) (Kırval and Çalışkan 2022). Levy on a bunker application will be introduced as a whole new measure. Still, each passing day is in favor of ETS in terms of adaptation to new policies on maritime transportation.

11.3.2 Possible Outcomes

Several submissions to open up the discussions have failed due to the opposition of many states. However, the recent MEPC meetings have shown that possible MBM for in-service vessels or marine bunkers is likely to be announced in 2023 (Psaraftis 2021; ICCT 2018; France 2018; Antigua and Barbuda 2018; United Kingdom 2020; The Marshall Islands 2020). Depending on the evaluations given, possible outcomes of the discussion can be summarized as follows:

(1) low-level levy with low-level ambitions to reduce emissions that may punish a continent's trade,
(2) matching the level of levy with ETS that can create harmonic issues and non-integrated approaches,
(3) a risky high-level levy that can cause market disruption in highly volatile market conditions,
(4) an ETS that can be a loosen-up version of EU ETS, which may be two times costlier in comparison with a levy,
(5) medium-level ETS that may create an imbalance of trade between different types of cargo trader countries,
(6) high-level ETS that the consequences of failure (voyage evasion and carbon leakage) of an applied and successful system for the EU being used to world shouldered by the LDCs and SIDS,
(7) rebate mechanism built on ETS that creates an extra administrative burden and long-lasting argumentation,
(8) business-as-usual scheme that may come after the postponed decision of an MBM, which possibly deescalating EU's decision with the monetary burden on EU operators,
(9) speed limits or speed optimization variants that can destroy a few branches of international shipping (i.e., perishable goods trade between South America and East Asia), and
(10) alternative fuel incentives without taxes that fail to generate GHG fund.

Among outcomes, each possible scheme has its own risk. Some are bearing lower impact with a high reward ratio.

11.4 Discussion

The interpretation of how MBMs perform in today's conditions is given in Table 11.3. The explanation for this situation is that ETSs have made progress in compliance lately, but the levy still has an overwhelming advantage.

As stated earlier, low-level levy derivatives offer neither vast improvements in fuel preference nor possible speed optimization incentives. Consequently, eliminating the first outcome even though it is favored is necessary to honor many criteria. The second outcome is one of very few concordant results. However, it can be predicted that consequences will push the European maritime trade to a disadvantageous position if it is applied together with the ETS. Harmonic problems are inevitable unless unity is achieved in practice. Although environmental organizations frequently emphasize the third possibility, it can be predicted that the trade environment will be damaged in the current state of fuel economics. Still, a feasibility study should be conducted to determine its affordability.

ETS is comparable or even ahead of the bunker levy in criteria of environmental effectiveness, practicality, relevancy to conventions, and compatibility with the existing policy framework. As a result of this situation, the comparison of the financial burden brought by the ETS should be scaled on the impact ratio. As long as the ETS remains low, it will be to the detriment of the European shipping industry and to the detriment of world shipping or the environmental approach at mid-range and above. The most reasonable decision would be the middle ground of the fourth and fifth possibilities regarding suitability. Most of these assumptions are not possible via direct EU ETS decision, but with the combination of the RM and ETS indicated in the seventh possibility. Otherwise, the fundamental dangers of the ETS are likely to be transferred to SIDS and LDCs.

From an environmental point of view, the eighth possibility can be called a loss of gains. The ninth and tenth possibilities, which were not discussed before in this article, are secondary returns of the two main MBMs discussed. They are directly targetable by IMO and states, according to their high environmental output provided by MACC

Table 11.3 Two MBM's comparative analyses on meeting with original nine criteria

Short representative names of criteria	ETS	Levy
Environmental effectiveness	X	X
Cost-effectiveness		X
Potential catalyzer effect		X
Practicality		X
Capacity building		X
Relevancy to conventions	X	
Administrative burden		X
Operational impact		X
Compatibility with existing policy framework	X	

analyses. It is helpful to separate the ninth possibility as optimization and limit. The limit is in a position to be criticized in many respects. Besides possible benefits, the threats to some maritime transport sectors make this possibility inapplicable. On the other hand, speed optimization looks like a system that needs to be handled separately for each ship and route. Unless its implementation has a low level of inclusiveness, it is possible to have a more complex nature than the ETS. The tenth possibility is weak for the same reasons as the eighth possibility.

As a result, out of ten possibilities, the maximum possible benefits would be the second possibility and the middle ground of the fourth and fifth possibilities.

11.5 Conclusion

This article evaluated MBMs in today's maritime and world conditions in the context of the original nine criteria. Possible applications were listed, and a comparison was made among them. As a result, the possibility of a low-to-medium-level ETS with a rebate mechanism and a medium-level levy outweighed other options based on the literature's reasonings. In line with the literature, the bunker levy could still be considered superior MBM in current conditions. Regarding economic effectiveness, a one-dimensional perspective may result in an appraisal of the ETS; however, being a catalyzer of new technologies, preventing offsetting, and practicality still belongs to the bunker levy. While legal aspects have benefited ETS over the past decade, the bunker levy bypasses these hurdles due to its core structure.

In the past decade, hesitant acts have increased the damage done by sea transportation to the world. Although one of the essential points for decarbonizing maritime transport seems to be the selection of the right MBM, both MBMs can be considered reasonable and recommended in terms of environmental benefits. Primarily, the issue of modal shifts is of vital importance when drafting the policy. Even though LDCs and SIDS are highly regarded in previous policy discussions, their vulnerability should still be subjected to positive discrimination. Such threats should be arranged in the first drafts, not subsequent amendments.

The most important of the future studies on this subject is how much offsetting of ETS will push the environmental recovery out of maritime transportation. It is also worth mentioning that prior to an important decision, a MACC study to be carried out in today's conditions is indispensable for selecting the MBM. Also, the level of voyage evasion can be estimated via the maritime transport network's response built on Liner Shipping Bilateral Connectivity Index. Subjects after a possible implementation that should be researched are anomaly analyses and difference maps of emissions via satellite imagery sensors.

References

Antigua and Barbuda (2018) Action plan for implementing the IMO GHG strategy and candidate measures, submitted by Antigua and Barbuda, Kenya, Marshall Islands, Palau, Solomon Islands, Tonga, and Tuvalu. IMO doc ISWG-GHG 4/2/3

Avetisyan M (2018) Impacts of global carbon pricing on international trade, modal choice and emissions from international transport. Energy Econ 76:532–548. https://doi.org/10.1016/j.eneco.2018.10.020

Beder S (2013) Environmental principles and policies: an interdisciplinary introduction. Routledge. Chapter 2. London. First published: 2006. E-ISBN: 9781315065908. https://doi.org/10.4324/9781315065908

BHP Group, Limited BW Group, DNB & DNV GL (2019) Carbon levy evaluation "could a carbon levy in shipping be an effective way to help reach the IMO greenhouse gas reduction goals". Global Maritime Forum: Singapore, 2019. globalmaritimeforum.org/content/2019/10/Could-a-carbon-levy-in-shipping-be-an-effective-way-to-help-reach-the-IMO-greenhouse-gas-reduction-goals-DNV-BW-BHP-DNB.pdf

Cariou P, Cheaitou A (2012) The effectiveness of a European speed limit versus an international bunker-levy to reduce CO_2 emissions from container shipping. Transp Res Part d: Transp Environ 17(2):116–123. https://doi.org/10.1016/j.trd.2011.10.003

Carlo R, Marc BJ, Santiago SD, Smith T, Søgaard K (2020) Aggregate investment for the decarbonization of the shipping industry. University Maritime Advisory Services-UMAS. Available at: https://www.globalmaritimeforum.org/content/2020/01/Aggregate-investment-for-the-decarbonisation-of-the-shipping-industry.pdf. Accessed 21 July 2022

Chai KH, Lee XN, Gaudin A (2019) a systems perspective to market–based mechanisms (MBM) Comparison for International Shipping. Ssrn Electron J

Christodoulou A, Dalaklis D, Ölcer A, Ballini F (2021) Can market-based measures stimulate investments in green technologies for the abatement of ghg emissions from shipping? A review of proposed market-based measures. Trans Maritime Sci 10(01):208–215. https://hrcak.srce.hr/258060

Cristea A, Hummels D, Puzzello L, Avetisyan M (2013) Trade and the greenhouse gas emissions from international freight transport. J Environ Econ Manag 65(1):153–173. https://doi.org/10.1016/j.jeem.2012.06.002

Denmark (2010) The International fund for greenhouse gas emissions from ships (GHG Fund) proposed by Cyprus, Denmark, the Marshall Islands, Nigeria and IPTA, IMO doc. MEPC 60/4/8

Devanney JW (2010) The impact of EEDI on VLCC design and CO_2 emissions. Center for Tankship Excellence. USA. http://www.c4tx.org/ctx/pub/eedi_athens.pdf

DNV GL (2021) MRV and DCS. Accessible at: https://www.dnv.com/maritime/insights/topics/MRV-and-DCS/index.html. Accessed 4 August 2022

Ellerman AD, Convery FJ, De Perthuis C (2010) Pricing carbon: the European Union emissions trading scheme. Cambridge University Press

European Commission (2021) Directive of the European Parliament and of the Council: amending Directive 2003/87/EC establishing a system for greenhouse gas emission allowance trading within the Union. Decision (EU) 2015/1814 concerning the establishment and operation of a market stab. 0211. 6. europarl.europa.eu/doceo/document/TRAN-AD-704649_EN.pdf

European Union (2019) Communication from the commission to the European Parliament, the European Council, the Council, the European Economic and Social Committee and the Committee of the Regions "the European Green Deal". Com/2019/640 Final. eur-lex.europa.eu/resource.html?uri=cellar:b828d165-1c22-11ea-8c1f-01aa75ed71a1.0002.02/DOC_1&format=PDF

Faber J, Markowska A, Nelissen D, Davidson M, Eyring V, Cionni I, Selstad E, Kågeson P, Lee D, Buhaug Ø, Lindtsad H, Roche P, Humpries E, Graichen J, Cames M, Schwarz W (2009) Technical support for European action to reducing Greenhouse Gas Emissions from international maritime transport. In: CE Delft, Vol 53, Issue 9

FOE (2009) A dangerous obsession. In: Foreign affairs, Vol 73, Issue 2.
FOE (2010) Clearing the air moving on from carbon trading to real climate solutions. http://etnar.net/wp-content/uploads/2014/03/clearing_air.pdf
France (2010) Further elements for the development of an emissions trading system (ETS) for International shipping proposal by France, submitted by France, IMO doc. MEPC 60/4/41
France (2018) Proposal to include work on market-based measures in the programme of follow-up actions of the initial IMO GHG strategy, submitted by France, IMO doc. ISWG-GHG 4/2/11
Germany (2010) Impact Assessment of an emissions trading scheme with a particular view on developing countries, IMO doc. MEPC 60/4/54
Gil-Lopez T, Verdu-Vazquez A (2021) Environmental analysis of the use of liquefied natural gas in maritime transport within the port environment. Sustainability 13(21):11989. https://doi.org/10.3390/su132111989
Giziakis C, Christodoulou A (2012) Environmental awareness and practice concerning maritime air emissions: the case of the Greek shipping industry. Marit Policy Manag 39(3):353–368. https://doi.org/10.1080/03088839.2012.671543
Gkonis KG, Psaraftis HA (2012) Modeling tankers' optimal speed and emissions. In: SNAME maritime convention. OnePetro. https://doi.org/10.5957/SMC-2012-A08
Gu Y, Wallace SW, Wang X (2019) Can an Emission Trading Scheme really reduce CO_2 emissions in the short term? Evidence from a maritime fleet composition and deployment model. Transp Res Part d: Transp Environ 74(August):318–338. https://doi.org/10.1016/j.trd.2019.08.009
Halim RA, Smith T, Englert D (2019) Understanding the economic impacts of greenhouse gas mitigation policies on shipping what is the state of the art of current modeling approaches? Executive summary for policy-makers in international maritime transport-world bank group report. http://documents.worldbank.org/curated/en/718061546900387225/
High-level Advisory Group (2009) High-level advisory group on climate change financing report work stream 2: paper on potential revenues from international maritime and aviation sector policy measures. https://www.cbd.int/financial/interdevinno/un-climate-transport.pdf
ICCT (2018) Policy update: the International Maritime Organization's Initial Greenhouse Gas Strategy, April 2018.https://theicct.org/sites/default/files/publications/IMO_GHG_StrategyFInalPolicyUpdate042318.pdf
ICS (2018) Reducing CO_2 Emissions to Zero: The 'Paris Agreement for Shipping'. Implementing the initial strategy on reduction of GHG Emissions from Ships.
IMO (2010) Full report of the work undertaken by the expert group on feasibility study and impact assessment of possible market-based measures, IMOdoc. MEPC 61/INF.2. http://martrans.org/documents/2009/air/MEPC61-INF2.pdf
IMO (2011) Inclusion of regulations on energy efficiency for ships in MARPOL Annex VI. IMOdoc. MEPC 62/24/Add. 1. Annex 19 Resolution MEPC.203(62) Adopted on 15 July 2011 https://wwwcdn.imo.org/localresources/en/OurWork/Environment/Documents/Technical%20and%20Operational%20Measures/Resolution%20MEPC.203(62).pdf
IMO (2018) Initial IMO strategy on reduction of GHG emissions from ships. Resolution MEPC.304(72) (adopted on 13 April 2018). IMO doc. MEPC 72/17/Add.1. Annex 11. imo.org/en/MediaCentre/HotTopics/Pages/Reducing-greenhouse-gas-emissions-from-ships.aspx
IMO (2019) Proposal to Establish an International Maritime Research and Development Board (IMRB). Submitted by BIMCO, CLIA, ICS, INTERCARGO, INTERFERRY, INTERTANKO, IPTA, and WSC; International Maritime Organization (IMO): London. UK. https://www.ics-shipping.org/wp-content/uploads/2019/02/proposal-to-establish-an-international-maritime-research-and-development-board-imrb-2-compressed-1.pdf
IMO (2021a) Fourth IMO GHG study. International Maritime Organisation (IMO) Faber J, Hanayama S, Zhang S, Pereda P, Comer B, Hauherhof H, van der Loeff W, Smith T, Zhang Y, Kosaka H, Adachi M, Bonello J, Galbarith C, Gong Z, Hirata K, Hummels D, Klejin A, Lee DS, Liu Y, Lucchesi A, Mao X, Muraoka E, Osipova L, Qian H, Rutherford D, Suarez de la Fuante S, Yuan H, Perico V, Wu L, Sun D, Yoo D, Xing H, 4 Albert Embankment.

London. SE1 7SR. https://wwwcdn.imo.org/localresources/en/OurWork/Environment/Documents/Fourth%20IMO%20GHG%20Study%202020%20Executive-Summary.pdf

IMO (2021b) Further shipping GHG emission reduction measures adopted. 17 June 2021b. https://www.imo.org/en/MediaCentre/PressBriefings/pages/MEPC76.aspx. Accessed 20 July 2022

IMO (2021c) Proposal for IMO to establish a universal mandatory greenhouse gas levy. submitted by the Marshall Islands and Solomon Islands. .Submitted by the Marshall Islands and the Solomon Islands, IMO doc. MEPC 76/7/12. https://wwwcdn.imo.org/localresources/en/OurWork/Environment/Documents/Air%20pollution/IMO%20submission%20to%20SBSTA%2052-55.pdf

International Union for the Conservation of Nature (2010) A rebate mechanism (RM) for a market-based instrument for international shipping proposal by IUCN, IMO doc. MEPC 60/4/55

Intergovernmental Panel on Climate Change (2021) Summary for policymakers. In: Masson-Delmotte V, Zhai P, Pirani A, Connors SL, Péan C, Berger S, Caud N, Chen Y, Goldfarb L, Gomis MI, Huang M, Leitzell K, Lonnoy E, Matthews JBR, Maycock TK, Waterfield T, Yelekçi O, Yu R, Zhou B (eds), Climate change 2021: the physical science basis. contribution of working group I to the sixth assessment report of the intergovernmental panel on climate change. Cambridge University Press, Cambridge, United Kingdom and New York, NY, USA, pp 3−32. https://doi.org/10.1017/9781009157896.001. https://www.ipcc.ch/report/ar6/wg1/

ITF/OECD (2018) Decarbonizing maritime transport. Pathways to zero-carbon shipping by 2035. In: International transport forum. https://www.itf-oecd.org/sites/default/files/docs/decarbonising-maritime-transport.pdf

Jamaica (2010) Achieving reduction in greenhouse gas emissions from ships through Port State arrangements utilizing the ship traffic, energy and environment model, STEEM (PSL) proposal by Jamaica, IMO doc. MEPC 60/4/40

Japan & WSC (2011) Consolidated proposal of "efficiency incentive scheme" based on the leveraged incentive scheme and the vessel efficiency system, IMO doc. GHG WG 3/3/2

Japan (2010) The leveraged incentive scheme (LIS) to improve the energy efficiency of ships based on the International GHG Fund proposed by Japan, IMO doc. MEPC 60/4/37

Joung TH, Kang SG, Lee JK, Ahn J (2020) The IMO initial strategy for reducing Greenhouse Gas (GHG) emissions, and its follow-up actions towards 2050. J Int Maritime Safety Environ Affairs and Shipping 4(1):1–7. https://doi.org/10.1080/25725084.2019.1707938

Kachi A, Mooldijk S, Warnecke C (2019) Carbon pricing options for international maritime emissions. New climate-institute for climate policy and global sustainability gGmbH: Berlin, Germany

Kågeson P (2011) The financial implications of a Levy & GHG Fund. http://urn.kb.se/resolve?urn=urn:nbn:se:vti:diva-646

Kapetanis GN, Gkonis KG, Psaraftis HN (2014) Estimating the operational effect of a bunker levy: the case of handymax bulk carriers. In Transport Research Arena 2014. https://orbit.dtu.dk/en/publications/estimating-the-operational-effect-of-a-bunker-levy-the-case-of-ha

Kırval L, Çalışkan UY (2022) Influence of the European Union (EU) on international maritime organization (IMO) about the market-based measures on emissions. Int J Environ Geoinformat (IJEGEO), 146–153. https://doi.org/10.30897/ijegeo.1047467

Kosmas V, Acciaro M (2017) Bunker levy schemes for greenhouse gas (GHG) emission reduction in international shipping. Transp Res Part d: Transp Environ 57:195–206. https://doi.org/10.1016/j.trd.2017.09.010

Lagouvardou S, Psaraftis HN, Zis T (2020) A literature survey on market-based measures for the decarbonization of shipping. Sustainability 12(10):3953. https://doi.org/10.3390/su12103953

Lema E, Karaganis A, Papageorgiou E (2017) A fuzzy logic modeling of measures addressing shipping CO_2 emissions. J Intell Syst 26(3):439–455. https://doi.org/10.1515/jisys-2015-0161

Maritime Magazine (2021) Maersk CEO calls for big carbon tax on bunker fuel. https://maritimemag.com/en/maersk-ceo-calls-for-big-carbon-tax-on-bunker-fuel/. Accessed 20 July 2022

Marine link (2022) Ship fuel and lube costs are Skyrocketing. https://www.marinelink.com/news/ship-fuel-lube-costs-skyrocketing-497237. Accessed 19 July 2022

Miola A, Marra M, Ciuffo B (2011) Designing a climate change policy for the international maritime transport sector: Market-based measures and technological options for global and regional policy actions. Energy Policy 39(9):5490–5498. https://doi.org/10.1016/J.ENPOL.2011.05.013

Mundaca G, Strand J, Young IR (2021) Carbon pricing of international transport fuels: impacts on carbon emissions and trade activity. J Environ Econ Manag, 102517. https://doi.org/10.1016/j.jeem.2021.102517

Norway (2010) The global emission trading system (ETS) for international shipping proposal by Norway, IMO doc. MEPC 61/4/22

Parry I, Heine D, Kizzier K, Smith T (2022) A carbon levy for international maritime fuels. Rev Environ Econ Policy 16(1):25–41. https://doi.org/10.1086/717961

Parry I, Heine MD, Kizzier K, Smith T (2018) Carbon taxation for international maritime fuels: assessing the options. International Monetary Fund. ISBN: 148437455X, 978-1-4843-7455-9. https://doi.org/10.5089/9781484374559.001

Paterson M (2021) COP26 agreed rules on trading carbon emissions—but they're fatally flawed. Published: 17/12/2021. Accessible at: https://theconversation.com/cop26-agreed-rules-on-trading-carbon-emissions-but-theyre-fatally-flawed-173922. Accessed 4 August 2022

Psaraftis HN (2012) Market-based measures for greenhouse gas emissions from ships: a review. WMU J Marit Aff 11(2):211–232. https://doi.org/10.1007/s13437-012-0030-5

Psaraftis HN (2019a) Speed optimization versus speed reduction: are speed limits better than a bunker levy? Maritime Econ Logist 21(4):524–542. https://doi.org/10.1057/s41278-019-00132-8

Psaraftis HN (2019b) Decarbonization of maritime transport: to be or not to be? Maritime Econ Logist 21(3):353–371. https://doi.org/10.1057/s41278-018-0098-8

Psaraftis HN (2021) Shipping decarbonization in the aftermath of MEPC 76. Cleaner Logist Supply Chain 1:100008. https://doi.org/10.1016/j.clscn.2021.100008

Psaraftis HN, Zis T, Lagouvardou S (2021) A comparative evaluation of market based measures for shipping decarbonization. Maritime Transp Res 2:100019. https://doi.org/10.1016/j.martra.2021.100019

Psaraftis HN (2022) Shipping Decarbonization: the MAC and MBM agendas. Blue sky maritime coalition- chartering commercial and finance Workstream. DTU Manag. backend.orbit.dtu.dk/ws/portalfiles/portal/273713837/Psaraftis_MAC_MBM_talk_08_04_2022_v2.0.pdf

Psaraftis HN, Lagouvardou S (2019) market based measures for the reduction of green house gas emissions from ships: a possible way forward. Samfundsoekonomen 2019(4):60–70. https://orbit.dtu.dk/en/publications/market-based-measures-for-the-reduction-of-green-house-gas-emissi

Shi Y (2016) Reducing greenhouse gas emissions from international shipping: Is it time to consider market-based measures? Marine Policy, 64https://doi.org/10.1016/j.marpol.2015.11.013

Shi Y, Gullett W (2018) International regulation on low-carbon shipping for climate change mitigation: development, challenges, and prospects. Ocean Dev Int Law 49(2):134–156. https://doi.org/10.1080/00908320.2018.1442178

Ship & Bunker (2022) Rotterdam bunker prices. https://shipandbunker.com/prices/emea/nwe/nl-rtm-rotterdam#LNG. Accessed 8 July 2022

Skjølsvik KO, Andersen AB, Corbett JJ, Skjelvik JM (2000) Study of greenhouse gas emissions from ships (MEPC 45/8 report to international maritime organization on the outcome of the IMO study on greenhouse gas emissions from ships), MARINTEK Sintef Group. MARINTEK Sintef Group, Trondheim, Norway

Tanaka H, Okada A (2019) Effects of market-based measures on a shipping company: using an optimal control approach for long-term modeling. Res Transp Econ 73:63–71. https://doi.org/10.1016/j.retrec.2019.01.006

The Marshall Islands (2020) Comments on the fourth IMO GHG study 2020 and encouraging further ambitious action to reduce GHG emissions. Submitted by the Marshall Islands and Solomon Islands, IMO doc. MEPC 75/7/17

The Bahamas (2010) Market-based instruments: a penalty on trade and development proposal by the Bahamas, IMO doc. MEPC 60/4/10
The United Kingdom (2010) Global emissions trading system (ETS) for international shipping proposal by the United Kingdom, IMO doc. MEPC 60/4/26
The United States of America (2010) The United States proposal to reduce greenhouse gas emissions from international shipping, the ship efficiency and credit trading (SECT), IMO doc. MEPC 60/4/12
Tillig F, Ringsberg JW, Psaraftis HN, Zis T (2020) Reduced environmental impact of marine transport through speed reduction and wind assisted propulsion. Transp Res Part d: Transp Environ 83:102380. https://doi.org/10.1016/j.trd.2020.102380
Tiwari S, Wee HM, Zhou Y, Tjoeng L (2021) Freight consolidation and containerization strategy under business as usual scenario & carbon tax regulation. J Clean Prod 279:123270. https://doi.org/10.1016/j.jclepro.2020.123270
Trafigura (2020) A proposal for an IMO-led global shipping industry decarbonization programme. Report by Trafigura. trafigura.com/brochure/a-proposal-for-an-imo-led-global-shipping-industry-decarbonisation-programme
Trivyza NL, Rentizelas A, Theotokatos G (2019) Impact of carbon pricing on the cruise ship energy systems optimal configuration. Energy 175:952–966. https://doi.org/10.1016/j.energy.2019.03.139
United States Congressional Budget Office (2008) CBO study on cap and trade (Issue February)
United Kingdom (2020) Economic incentives to reduce GHG emissions from international shipping submitted by the United Kingdom, IMO doc ISWG-GHG-7/8/1
United Nations Conference on Environment and Development (1992) Report of the united nations conference on environment and development annex I: rio declaration on environment and development. Rio de Janeiro, Brazil, 3–14 June 1992. un.org/en/development/desa/population/migration/generalassembly/docs/globalcompact/A_CONF.151_26_Vol.I_Declaration.pdf
Wang C, Chen J (2017) Strategies of refueling, sailing speed and ship deployment of containerships in the low-carbon background. Comput Ind Eng 114:142–150. https://doi.org/10.1016/j.cie.2017.10.012
Wang C, Xu C (2015) Sailing speed optimization in voyage chartering ship considering different carbon emissions taxation. Comput Ind Eng 89:108–115. https://doi.org/10.1016/j.cie.2015.04.034
Wang X, Norstad I, Fagerholt K, Christiansen M (2019) Green tramp shipping routing and scheduling: effects of market-based measures on CO_2 reduction. In: Sustainable shipping, pp 285–305. Springer, Cham. https://doi.org/10.1007/978-3-030-04330-8_8
World Bank, & Ecofys (2018) State and trends of carbon pricing 2018. In Zeitschrift für Orthopädie und ihre Grenzgebiete (Vol 137, Issue 04). https://doi.org/10.1055/s-2008-1037045
WSC (2010) Vessel efficiency system (VES) proposal by World Shipping Council, IMO doc. MEPC 60/4/39

The manufacturer's authorised representative in the EU is Springer Nature Customer Service Centre GmbH, Europaplatz 3, 69115 Heidelberg, Germany. If you have any concerns regarding our products, please contact ProductSafety@springernature.com

Printed and bound by CPI Group (UK) Ltd, Croydon, CR0 4YY
30/09/2026
02225303-0003